Subsea Robotics

Subsea Robotics addresses the complex challenges in underwater oceanographic data collection, submerged infrastructure inspection, and subsea defence missions. This book offers a comprehensive, system-level perspective on subsea robotics by integrating engineering fundamentals with real-world deployment practices. It covers design, propulsion, autonomy, navigation, communication, and operational standards. By combining core engineering principles with deployment realities, the book shows how subsea uncrewed systems are designed and built to meet real-world constraints.

FEATURES

- Explores interdisciplinary integration of systems for Unmanned Underwater Vehicles (UUVs)
- Explains autonomy architecture, control architecture, and human–machine interfaces for UUVs
- Shares practical experiences from UUV operations and examples of underwater survey missions
- Discusses energy and propulsion options, requirements, limitations, and key design factors
- Includes applications, case studies, missions, and examples illustrating the development of UUVs

This book is intended for interdisciplinary researchers, professional engineers and graduate students in ocean engineering, naval architecture, robotics, and electrical engineering.

Subsea Robotics

From Design to Deployment

Dr. Amit Ray

CRC Press
Taylor & Francis Group
Boca Raton London New York

CRC Press is an imprint of the
Taylor & Francis Group, an **informa** business

Designed cover image: © Amit Ray

First edition published 2027
by CRC Press
2385 NW Executive Center Drive, Suite 320, Boca Raton FL 33431

and by CRC Press
4 Park Square, Milton Park, Abingdon, Oxon, OX14 4RN

CRC Press is an imprint of Taylor & Francis Group, LLC

ISBN: 9781041233817 (hbk)
ISBN: 9781041235064 (pbk)
ISBN: 9781003737476 (ebk)

DOI: 10.1201/9781003737476

Typeset in Times New Roman
by Deanta Global Publishing Services, Chennai, India

For my parents, Jaya and Kalyan

Contents

Abbreviations

6-DOF	Six degrees-of-freedom
ABS	American Bureau of Shipping
ACOMMS	Acoustic communications
ADCP	Acoustic Doppler Current Profiler
AI	Artificial Intelligence
ASME	American Society of Mechanical Engineers
ASTM	American Society for Testing and Materials
ASV	Autonomous Surface Vehicle (or Vessel)
ASW	Anti-Submarine Warfare
AUV	Autonomous Underwater Vehicle
BLDC	Brushless Direct Current (motor)
BMS	Battery management system
C4	Command, control, communication, and computers
CAN	Controller Area Network
CFD	Computational fluid dynamics
CFRP	Carbon fibre reinforced plastic
CMOS	Complementary metal-oxide-semiconductor
CONOPS	Concept of Operations
CSIR	Council for Scientific and Industrial Research
CTD	Conductivity–Temperature–Depth (sensor)
CURV	Cable-Controlled Underwater Recovery Vehicle
DARPA	Defense Advanced Research Projects Agency
DDS	Data Distribution Service
DNV-GL	Det Norske Veritas-Germanischer Lloyd
DoD	Department of Defense (US)
DRDO	Defence Research and Development Organisation (MoD, India)
DVL	Doppler velocity log
EKF	Extended Kalman Filter
ELF	Extremely low frequency (electromagnetic waves)
ET&E	Experimental Test & Evaluation
FOG	Fibre optic gyro
GNC	Guidance, navigation, and control
GoI	Government of India
GPS	Global positioning system
GRP	Glass reinforced plastic
GUI	Graphical user interface
HALE	High Altitude Long Endurance (UAV)
HAUV	Hybrid aerial underwater vehicle
HIL	Hardware-in-loop (testing or simulation)
HMI	Human–machine interaction
H-ROV	Hybrid ROV

I-AUV	Intervention-AUV
iDEX	Innovations for Defence Excellence
IEEE	Institute of Electrical and Electronics Engineers
IMU	Inertial measurement unit
INU	Inertial navigation unit
IoT	Internet of Things
IoUT	Internet of Underwater Things
ISO	International Organization for Standardization
ISR	Intelligence Surveillance and Reconnaissance
IV&V	Integration, verification, and validation
JAUS	Joint Architecture for Unmanned Systems
KF	Kalman Filter
LARS	Launch and recovery system
LBL	Long baseline (acoustic communication)
LDUUV	Large displacement unmanned underwater vehicle
LiDAR	Light detection and ranging
Li-ion	Lithium-Ion (cell chemistry)
LiPo	Lithium-Polymer
LOS	Line-of-sight (communication)
MALE	Medium Altitude Long Endurance (UAV)
MBES	Multi-beam echosounder
MCM	Mine Countermeasure (operations)
MEMS	Micro electro-mechanical systems
ML	Machine Learning
MoD	Ministry of Defence (GoI)
MoES	Ministry of Earth Sciences (GoI)
MOOS-IvP	Mission-Oriented Operating Suite—Interval Programming
MiMH	Nickel-Metal Hydride (cell chemistry)
MSME	Micro, small, and medium enterprises
NIO	National Institute of Oceanography (Goa)
NIOT	National Institute of Ocean Technology (Chennai)
NMEA	National Marine Electronics Association
NOAA	National Oceanic and Atmospheric Administration
NSTL	Naval Science & Technological Laboratory (DRDO, India)
ONR	Office of Naval Research
OTEC	Ocean thermal energy conversion
PEM	Proton Exchange (or Polymer Electrolyte) Membrane (Fuel Cell)
PLAN	People's Liberation Army Navy (China)
PLUSNet	Persistent Littoral Undersea Network
RF	Radio frequency
ROS	Robot Operating System
ROV	Remotely Operated Vehicle
SAS	Synthetic Aperture Sonar
SATCOM	Satellite communication
SBL	Short baseline (acoustic navigation)

SE	Systems engineering
SLAM	Simultaneous localization and mapping
SIoT	Subsea Internet of Things
SNR	Signal-to-noise ratio
SPAWAR	Space and Naval Warfare Systems Command (US Navy)
SSS	Side scan sonar
TTL&R	Torpedo tube launch and recovery
UAV	Unmanned (or uncrewed) aerial vehicle
UHR	Underwater Humanoid Robots
UKF	Unscented Kalman Filter
UMAA	Unmanned Maritime Autonomy Architecture
UMCC	Unmanned Maritime Common Control
USBL	Ultra-short baseline (acoustic navigation)
USV	Unmanned surface vehicle
UUS	Unmanned underwater systems
UUV	Unmanned underwater vehicle
UWSN	Underwater wireless sensor network
UxS	Unmanned Systems (in general)
UXV	Unmanned surface/underwater vehicle
VCB	Vertical centre of buoyancy
VCG	Vertical centre of gravity
VLF	Very low frequency (electromagnetic waves)
WHOI	Woods Hole Oceanographic Institution (US)
WPT	Wireless power transfer
XLUUV	Extra-large unmanned underwater vehicle

Preface

The ocean depths are vital for our planet's climate, marine ecosystem, and energy resources. Among the challenges of subsea engineering are the crushing hydrostatic pressure, the imperviousness of light or any electromagnetic waves, and the corrosive and dynamic environment. These offer opportunities for the multidisciplinary field of subsea robotics, which promises greater access to the ocean depths through remote or autonomous means.

My engagement with underwater robotics began in 2004 at IIT Kharagpur, where my MTech guide Professor Debabrata Sen and his colleague Professor CS Kumar were involved with the development of India's earliest indigenous AUVs. I was fortunate to witness AUV development at DRDO's NSTL since 2005, thanks to scientists such as Cmde N. Banerjee, Dr Manu Korulla, Dr HN Das, Dr Bharatwaja S. Aiyangar, Cmde PR Kulkarni, Dr Kushal Bhattacharya, and Dr Shiju John.

By 2010, after my PhD in Manoeuvring Hydrodynamics, I tried to drum up interest in underwater gliders as a long-endurance means for undersea data collection. Research projects did take off at NCW IIT Delhi, but it took another decade before aerial drones captured popular imagination. Only thereafter did unmanned (uncrewed) systems start appearing on the wish list of most navies.

In-house design of a large AUV was an ambitious project assigned to our design organisation in 2021. It was due to Admiral KB Singh's confidence in us and the dynamic leadership of VAdm Kiran Deshmukh and RAdm Nirmal Menon that our small team of submarine designers took the leap into the unmanned domain.

There are multiple facets of UUV design and development: mechanical, electronics, controls, programming, energy, propulsion, and naval architecture. It was a steep learning curve as we debated operational concepts, user requirements, fished for suitable equipment, and sought out experts. We named our AUV design 'Jalkapi', and the journey of its creation took my energetic team of Shashank Shankar, Alok Sahu, and Venkatesh Prabhu to every corner of the country. We were privileged to develop warm links with energetic individuals at Technology Innovation Hubs, IITs, CSIR, MoES, and DRDO labs, and of course start-ups.

Subsea or marine robotics requires convergence of several disciplines of engineering and an understanding of scientific and industrial/military applications. A fantastic experience of learning and networking with technologists, scientists, and professionals was the 'Marine Robotics School', a week-long workshop conducted by the Marine Instrumentation Division of the National Institute of Oceanography (NIO), Goa, in November 2023. I am deeply grateful to Dr Pramod Maurya and his colleagues for their guidance and for successful development of our test-bed AUV 'Jaya'.

Trials of AUVs are exciting, and our team of designers witnessed several. We dived in the sea off Dwarka with NIO's ROV. We participated in various AUV lake and harbour trials in Bengaluru, Kolkata, Chennai, Kochi, and Goa. Our small team of AUV designers and developers is now led by Pankaj Grover and includes Rohit Shekhar, Selva Muthukumaran, Manoj Kaimal, Rahul Krishna, and Kapil Yadav. I am deeply grateful to all of them, as well as to my family, friends, and my colleagues at DND(SDG), particularly RAdm SL Saji for his guidance and unflinching support.

The experiences of this journey motivated me to prepare an introductory text covering AUV/ROV applications, technologies, development roadmaps, and product examples. The National Maritime Foundation's award of the Admiral AK Chatterji Memorial fellowship in 2024 resulted in the book titled *Underwater Robotics: Engineering and Applications*. It included glimpses of our development journey and was oriented towards the Indian ecosystem of underwater technologies.

This book is focused more on the engineering aspects of designing, integrating, testing, and deploying unmanned underwater vehicles. It describes AUVs as well as ROVs (including manipulators), covering their design process, equipment, and technologies, but without mathematical or control theory rigour. While drawing upon Indian examples, the book highlights the state-of-the-art in various domains of subsea robotics and emerging trends for the future. I hope it will be found useful by students as well as industry practitioners, seeking familiarity with a non-core discipline, or exploring technology solutions for various aspects of AUVs and ROVs, from their design to deployment.

Amit Ray
New Delhi, 2026

About the Author

Amit Ray is a Naval Architect serving in the Indian Navy, presently as Commodore Naval Design (Submarine Design Group). He has been engaged in the design, construction, repairs, and trials of warships and submarines in the Navy for more than 25 years.

His awards include the IIT Kharagpur Director's Gold Medal for M.Tech (Ocean Engineering and Naval Architecture) in 2006 and the 'Medal of Distinction' of the Royal Institution of Naval Architects, UK (2010). His academic qualifications include M.Phil. (Defence and Strategic Studies) from Mumbai University and PhD in Submarine Hydrodynamics from IIT Delhi, India (2010). He has authored more than 30 technical papers and is a Fellow of the Institution of Naval Architects of India. He has been involved in the design and development of autonomous underwater systems and related technologies with various Indian institutions and start-ups.

His first book, *Tracing the Undersea Dragon: Chinese SSBN Programme and the Indo-Pacific*, was published in 2021. His second book, *Underwater Robotics: Engineering and Applications*, was published in 2025.

1 Introducing Subsea Robotics

1.1 INTRODUCTION

Covering over 70% of Earth's surface, the world's oceans represent the planet's largest and most enigmatic frontier. The profound physical diversity of subsea landscapes, from trenches and hydrothermal vents to vast abyssal plains, remains largely unexplored owing to the challenges posed by the deep-sea environment. Unlike terrestrial and aerial domains, underwater space is opaque to electromagnetic signals, imposes harsh pressure with increase in depth, and fundamentally constrains human presence.

Unmanned (or Uncrewed) underwater vehicles (UUVs) include autonomous underwater vehicles (AUVs) and Remotely Operated Vehicles (ROVs). Together, these vehicles and associated systems for operating underwater are referred to here as 'Subsea Robotics'. UUVs have brought transformative capabilities to ocean exploration and exploitation. These platforms multiply human reach across scientific, industrial, and defence domains, enabled by advances in miniaturised electronics, materials science, computing capability, and machine learning/artificial intelligence.

Autonomous and remotely operated underwater vehicles can play distinct as well as complementary roles in scientific exploration and industrial applications. Beginning with clues provided by towed or lowered instruments, autonomous underwater vehicles (AUVs) can localise and make preliminary photographic surveys, as well as fine-scale bathymetric mapping. Remotely operated vehicles (ROVs) enable close-up inspection, photo-mosaicing, and tasks involving the manipulation of samples and instruments. ROVs are the preferred means for underwater offshore structural survey or intervention (referred to as 'Work-class ROVs' when they undertake manipulation using robotic arms), and are also used to conduct seafloor tasks for scientific studies (Yoerger et al., 2007).

UUVs, together with their surface counterparts (USVs), form the broader domain of 'marine robotics' (Choi & Yuh, 2016). Collectively, USVs and UUVs are sometimes also referred to as UXVs, where X denotes either 'S' (Surface) or 'U' (Underwater). The fast-growing field of marine robotics is reshaping our engagement with oceans in several ways, making them more accessible, observable, and understandable (Griffiths, 2003).

This chapter outlines the historical evolution of UUVs and the challenges of underwater operations. The various categories of UUVs are mentioned, and their main components are outlined. This is followed by an overview of scientific, industrial, and defence applications of underwater or subsea robots.

DOI: 10.1201/9781003737476-1

1.2 EVOLUTION

Since the mid-20th century, underwater vehicle technology has evolved significantly. Innovation in the 1950s focused on manned submersibles for scientific research (Busby, 1976), requiring life-support systems to protect human occupants (Weir, 2001). The maiden dive to the Challenger Deep in the Pacific Ocean by the *Trieste* in 1960 and the development of the submersible *Alvin* by the Woods Hole Oceanographic Institution (WHOI) exemplify the innovations of that period (Ballard & Drew, 2021). By 1970, the total number of manned submersibles had increased dramatically, with over 100 built during the 1960s alone (Geyer, 1977).

However, the operational risk, cost, and limited endurance of manned submersibles motivated interest in uncrewed platforms (Sweeney, 1970). The 1950s–1960s witnessed the emergence of basic towed systems and tethered underwater vehicles—precursors to modern ROVs.

The growth of the offshore oil and natural gas industry after the 1950s shaped underwater technology and subsea engineering. The imperatives of offshore oil and natural gas extraction from seawater hundreds of metres deep led to new engineering solutions, which further increased to thousands of metres of water depth by the 1980s. This growth of offshore oil and gas industries catalysed the development of mechanised inspection and intervention systems, favouring ROVs over saturation divers due to safety and efficiency gains. 'Work-class' ROVs, equipped with manipulators, probes, cameras, and lights, were controlled by shipboard operators through the connecting umbilical that provides power and exchanges signals and data (Christ, 2011).

Subsequently, fully autonomous vehicles gained prominence from the late 1980s onward, with oceanographic research institutions pioneering early AUVs capable of survey and mapping missions. This technological foundation has enabled broader commercial, scientific, and military applications in the past two to three decades.

In the past two decades, the steep increase in cheap computing power, open source frameworks and affordable components has opened up the world of robotics for young persons, hobbyists, and small firms (Moore, Bohm, & Jensen, 2010). Innovations in the scientific and industrial domains are encouraging developments in the defence sector as well.

The growth of AI/ML solutions promises greater capabilities in autonomous data processing, edge computing, and coordinated functioning of underwater robots in the near future (Aoki, 2008). The ability to collect underwater data through surveys is the most important application of underwater robotics, which is common across civil and military domains.

1.3 SUBSEA SURVEY

Data collected from underwater surveys by AUVs are invaluable for a wide range of applications. This section describes how survey data are used across various domains and the purposes they serve.

1.3.1 Oceanography and Climate Studies

- **Bathymetric Mapping**: Detailed maps of the seafloor help scientists understand ocean circulation patterns, sediment transport, and tectonic activity.
- **Environmental Monitoring**: Data on water temperature, salinity, dissolved oxygen, and chlorophyll concentration provide insights into climate change, marine ecosystems, and biogeochemical cycles.
- **Sub-Bottom Profiling**: Investigating subsurface structures aids in understanding sedimentary processes, earthquake hazards, and the history of ocean basins.

1.3.2 Marine Biology and Ecology

- **Habitat Mapping**: High-resolution imagery and sonar data enable the mapping of coral reefs, seagrass beds, and other critical habitats.
- **Biodiversity Studies**: Surveys can help document species distribution and abundance, supporting conservation and management efforts.
- **Impact Assessments**: Monitoring environmental changes following events such as oil spills or coral bleaching provides critical data for environmental impact assessments.

1.3.3 Offshore Oil and Gas Exploration

- **Seafloor Characterisation**: Detailed bathymetric and sub-bottom data help identify potential drilling sites, assess reservoir properties, and plan pipeline routes.
- **Risk Assessment**: Surveys can identify underwater hazards such as shallow areas, rocky outcrops, or areas of high sedimentation that may pose risks to infrastructure.
- **Regulatory Compliance**: Accurate survey data are often required to meet regulatory standards and ensure safe installation of offshore infrastructure.

1.3.4 Marine Infrastructure Inspection

- **Pipeline and Cable Inspection**: AUVs are used to inspect underwater pipelines, cables, and other infrastructure, detecting corrosion, damage, or biofouling.
- **Port and Harbour Surveys**: Detailed surveys of harbour areas support dredging operations, maintenance planning, and navigational safety.
- **Renewable Energy**: In the context of offshore wind farms and tidal energy installations, AUV surveys help assess seabed conditions and monitor structural integrity.

1.3.5 Societal and Environmental Benefits

- **Disaster Response**: In the aftermath of natural disasters, underwater surveys can assess damage to coastal structures and provide data for reconstruction planning.
- **Archaeological Discoveries**: AUVs have led to significant discoveries of shipwrecks and submerged cultural heritage sites, enhancing our understanding of history.
- **Resource Management**: Data from underwater surveys support fisheries management, marine spatial planning, and environmental conservation initiatives.

1.4 CHALLENGES IN SUBSEA OPERATIONS

The underwater environment presents significant challenges for AUVs. Limited online communication due to the impossibility of underwater electromagnetic transmission and generally poor visibility for vision-based systems are major factors. Hydrodynamic effects strongly affect the nonlinear dynamics of actuators and the vehicle itself. Navigation is challenging, power sources are limited, and reliability is crucial, especially in dangerous environments or for long missions. Achieving a high degree of autonomy is difficult due to the highly hazardous and unknown environment, non-uniform and unstructured seawater, and the nonlinear response of vehicles.

These challenges can be broadly classified into environmental, technical, and operational categories and listed here. These will point us towards the features, capabilities, and limitations in the design of AUVs.

1.4.1 Environmental Challenges

- **Optical Attenuation**: Water attenuates light, reducing visibility and affecting the performance of optical cameras. Turbid waters or those with high plankton concentrations further reduce image quality.
- **Acoustic Variability**: Sound speed in water varies with temperature, salinity, and depth. These variations affect the performance of sonar systems and complicate data interpretation.
- **Pressure and Temperature Extremes**: At great depths, high pressure and low temperatures can affect both sensor performance and the structural integrity of the AUV. Equipment must be designed to withstand these harsh conditions, which increases cost and complexity.
- **Currents and Turbulence**: Ocean currents and turbulent flows can displace AUVs, affecting navigation accuracy and survey repeatability. Dynamic environments require advanced control algorithms and robust INS (inertial navigation systems) to maintain course.
- **Marine Life and Debris**: Biological entities and floating debris can interfere with sensors, cause collisions, or generate false returns from sonar.

1.4.2 Technical Challenges

1.4.2.1 Navigation and Positioning

Accurate navigation is paramount for geo-referencing survey data:

- **GPS Limitations**: Global Positioning System (GPS) signals do not penetrate water, so AUVs must rely on INS, acoustic positioning systems (such as USBL or LBL), and dead-reckoning, which can accumulate errors over time.
- **Sensor Drift**: Inertial sensors can drift over long missions, necessitating periodic calibration or correction from external references.
- **Terrain Referencing**: While terrain-aided navigation can improve accuracy, it requires preexisting high-quality bathymetric maps, which may not always be available.

1.4.2.2 Energy Management

- **Limited Power**: AUVs are typically battery-powered, and energy must be conserved for propulsion, sensor operation, data processing, and communication. Energy limitations restrict mission duration and the amount of data that can be transmitted.
- **Balancing Energy and Data Needs**: Energy conservation strategies must be balanced with the need for timely data reporting, posing a significant design challenge.

1.4.2.3 Communication Constraints

- **Bandwidth Limitations**: Underwater acoustic communication offers limited bandwidth and high latency, making real-time data transmission challenging.
- **Signal Attenuation and Multipath**: Acoustic signals are susceptible to multipath effects and attenuation, which can lead to data loss or corruption.
- **Intermittent Connectivity**: Communication links may be intermittent, necessitating robust on-board data storage and delayed data transmission strategies.

1.4.3 Operational and Logistical Challenges

1.4.3.1 Mission Planning and Execution

- **Complex Mission Planning**: Designing survey missions that maximise area coverage while ensuring high-resolution data collection requires detailed planning and sophisticated software tools.
- **Risk Management**: Operating AUVs in challenging underwater environments involves risks such as collisions, entanglement, or sensor malfunctions. Effective risk management protocols and redundancy in sensor systems are essential.

- **Recovery and Maintenance**: Post-mission recovery of AUVs can be challenging, especially in rough seas or remote areas. Additionally, the maintenance and repair of sophisticated AUV systems are costly and require specialised expertise.

1.4.3.2 Data Overload and Analysis

- **Volume of Data**: Modern AUVs generate massive amounts of data, which can overwhelm storage systems and make real-time analysis difficult.
- **Data Integration**: Integrating heterogeneous data streams from multiple sensors into a coherent dataset for analysis requires advanced data fusion and processing techniques.
- **Quality Control**: Ensuring data quality—such as correcting for sensor errors, filtering noise, and calibrating measurements—is an ongoing challenge.

1.5 CATEGORIES OF SUBSEA ROBOTS

1.5.1 Remotely Operated Vehicles (ROVs)

ROVs remain connected to a surface control station via a tether supplying power, control signals, and data. This arrangement permits complex manipulation, high-bandwidth real-time control, and sensor data streaming, vital to subsea construction, maintenance, and intervention tasks. They have features and capabilities to undertake tasks such as gripping, turning valves, and undertaking maintenance of subsea offshore structures (Christ, Wernli, & Wernli Sr, 2011). Beyond a depth of about 300 metres reachable by human divers, ROVs are essential for accessing complex equipment such as wellheads for drilling oil. An ROV is controlled by human operators from the mother ship or offshore platform for undertaking these tasks (Figure 1.1).

ROVs are categorised based on their functions into Classes I to V, generally in increasing order of size (American Bureau of Shipping, 2025), as follows:

- **Observation Only**: Primarily employed for underwater video observation. In general, they are small vehicles fitted with video cameras, lights, and thrusters.
- **Observation with Payload**: Vehicles capable of carrying additional sensors (over and above those carried by 'Pure Observation' ROVs or HROVs), such as data measurement instrumentation, sonar systems, additional video cameras, still cameras, etc.
- **Work Class**: Work-class ROVs or HROVs are primarily employed for conducting work while submerged. In general, these are larger and more powerful than observation ROVs. They commonly have multiplexing capability that allows additional sensors and tools to be added to the ROV or HROV and operated without the need for being separately hardwired through the umbilical.

FIGURE 1.1 ROV 'ROSUB-6000' of NIOT Chennai.

Source: **Author.**

- **Seabed-working vehicles**: These generally operate on the seafloor, where they travel/manoeuvre using wheels, belt traction systems, thrusters, water jets, or by combinations of these methods. These vehicles are configured for special-purpose tasks such as cable and pipeline trenching, excavation, dredging, or other seafloor construction work.

1.5.2 Autonomous Underwater Vehicles (AUVs)

Autonomy is essential where tether management from a mother ship is not practical, such as for deep ocean surveys, extended environmental monitoring, and covert military reconnaissance. AUVs operate without tethers and execute preprogrammed and adaptive missions.

FIGURE 1.2 AUV developed by CSIR-NIO, Goa.

Source: **Author.**

Their form is designed to minimise drag to enable longer transit to locations of interest (Figure 1.2). Energy storage limitations necessitate efficient propulsion and power management strategies. On-board health monitoring and decision-making requirements distinguish AUVs from ROVs, which have the benefit of continuous human monitoring and control. AUVs will be the major focus of this book.

The ABS Rules for Building and Classing Underwater Vehicles, Systems and Hyperbaric Facilities categorise AUVs and HAUVs as follows, in terms of external diameter (or maximum cross-sectional dimension):

- **Micro**: Diameter ≤ 3-inch (approx 8 cm), man-portable, launched from surface or submarine.
- **Small**: Diameter > 3-inch, and ≤ 10-inch (i.e., about 8 to 25 cm), man-portable, launched from surface or submarine.
- **Medium**: Diameter > 10-inch, and ≤ 21-inch (i.e., about 25 to 53 cm), launched from surface or submarine.
- **Large**: Diameter > 21-inch, and ≤ 84-inch (i.e., about 53 to 200 cm), launched from surface or submarine.
- **Extra Large**: Diameter > 84-inch (i.e., > 200 cm), launched from a pier.

1.5.3 Underwater Gliders

These are energy-efficient AUV variants utilising buoyancy-driven propulsion for long-term profiling missions (Davis, Eriksen, & Jones, 2002). They follow saw-tooth trajectories across ocean depths at slow speeds, consuming very little energy, recording temperature and conductivity data as they move. They communicate periodically via satellite when on the surface, transmitting information. Once launched from a mother ship (Figure 1.3), Gliders have successfully undertaken scientific missions of

FIGURE 1.3 'Seaglider' unmanned underwater vessel.

Source: **US Navy photo by Mass Communication Specialist 2nd Class Lily Daniels.**

several months' duration, displaying potential for networks for ocean surveillance (Ray, Singh, & Seshadri, 29–30 June 2011).

1.5.4 Hybrid Vehicles (H-AUVs or H-ROVs)

These refer to UUVs capable of both autonomous survey and active manipulation that have emerged by integration of AUV and ROV capabilities, such as the 'Aquanaut' shown in Figure 1.4. 'Hybrid' AUVs can also refer to designs that combine features of buoyancy-driven propulsion for low speeds and intermittent use of a thruster for higher speeds or better directional control.

1.6 MAIN COMPONENTS

In general, an uncrewed marine vehicle would have the following categories of systems:

- **Physical Systems**: Components of the vehicle that are essential for its useful functioning (propulsion, manoeuvring, sensors, manipulators, etc).
- **Command and Control Systems**: These are software, actuators, and interfaces required for vehicle self-health monitoring, decision-making, mission execution, and communication.

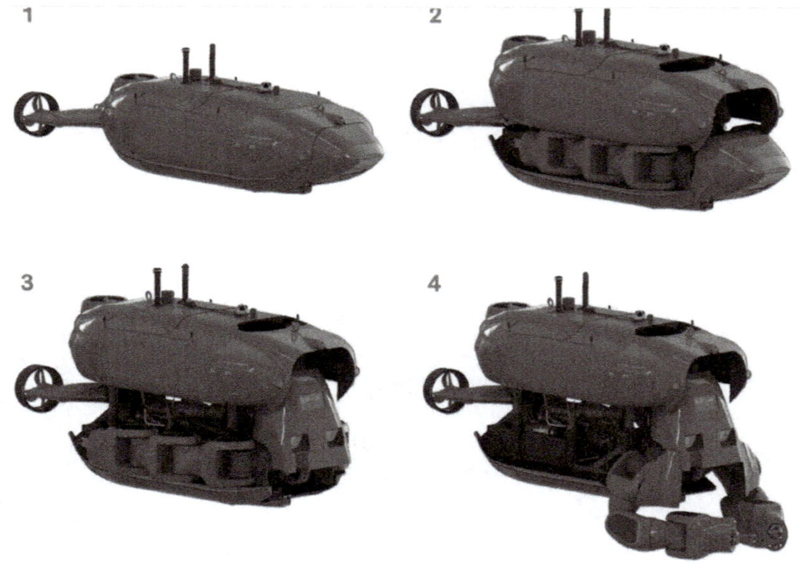

FIGURE 1.4 Aquanaut UUV by Houston Mechatronics (now Nautica Robotics).

Source: **Ackerman, 2019.**

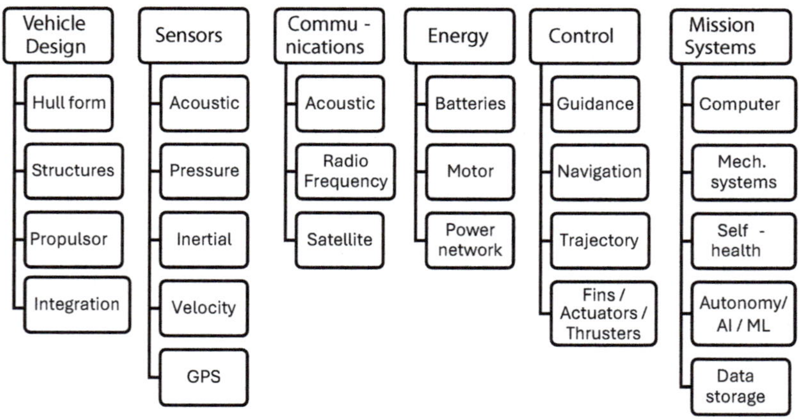

FIGURE 1.5 Systems and technologies in uncrewed marine vehicles.

Source: **Author.**

- **Support Systems**: These are facilities located on the support ship or ashore, essential for launch, recovery, maintenance, recharging, human monitoring, etc.

Effective underwater vehicles integrate multiple systems to overcome environmental challenges (Figure 1.5). These include structural design, hydrodynamics, control systems, energy storage and distribution systems, mechanical systems, navigation, communication, and overall mission control (autonomous operation). The major physical systems can be categorised as given below, which will be elaborated further in subsequent chapters.

1.6.1 Structural Design and Materials

Pressure vessels and load-bearing frames utilise high-strength aluminium, titanium, and composites engineered for hydrostatic loads and corrosion resistance. Modular construction enables efficient maintenance and mission adaptability.

1.6.2 Propulsion and Manoeuvring

Thruster designs optimise power-to-weight ratios and nozzle configurations to balance speed, noise reduction, and manoeuvrability. AUVs often incorporate buoyancy-driven propulsion to maximise endurance.

1.6.3 Sensors and Payload

Robust suites encompass inertial navigation systems (INS), Doppler velocity logs (DVL), sonar mapping (multibeam, side-scan), optical cameras, chemical sensors, and environmental probes (Griffiths, 2003).

1.6.4 Power Systems

Energy density and management govern mission duration. Lithium-ion batteries dominate, supplemented by emerging energy harvesting methods (e.g., thermal gradients, currents).

1.7 SCIENTIFIC USES OF UUVS

Unmanned underwater vehicles have become indispensable tools in scientific exploration, offering capabilities that have transformed oceanography, marine biology, environmental science, and underwater archaeology. Examples of AUVs for scientific applications are given in Table 1.1.

1.7.1 Oceanographic Research

Oceanography is the field that has benefited most from the advent of AUVs and UUVs (Figure 1.6). Their ability to collect high-resolution data over extended periods and in inaccessible areas has enabled scientists to undertake the following:

TABLE 1.1
Examples of AUVs Developed for Scientific Applications

AUV Example	Application Domain	Specific Applications
Autosub	Ocean science	Multidisciplinary research, ice-covered areas, terrain following, physical, geophysical, biological, chemical oceanography
MARES	Environment monitoring	Monitoring sea outfalls discharges, wastewater dilution mapping, modular design
HUGIN	Ocean science	Cold-water coral reef environmental surveys, fisheries stock assessment
Urashima	Ocean science	Long-distance cruising, sea bottom surveys in earthquake areas, global warming research, Arctic ice surveys
SeaBED	Marine geology	Studying large gas blowout structures
Slocum	Oceanography	Water column measurements, long endurance/range
Seaglider	Oceanography	Water column measurements, long endurance/range
Spray	Oceanography	Water column measurements, long endurance/range

FIGURE 1.6 Aster AUVs of Ifremer for scientific reconnaissance up to 2,850 metres depth.

Source: **Olivier Dugornay, https://upload.wikimedia.org/wikipedia/commons/6/6e/AUV_AsterX_et_navire_ oc%C3%A9anographique_L%27Europe_%28Ifremer_00560-67193_-_21929%29.jpg**

- **Map the Seafloor**: Advanced multibeam and synthetic-aperture sonar systems allow detailed bathymetric mapping. For example, AUVs have been deployed to map deep ocean trenches and mid-ocean ridges (Yoerger et al., 2007).
- **Study Water Column Properties**: Instruments such as CTD sensors, fluorometers, and dissolved oxygen meters are routinely mounted on UUVs to measure temperature, salinity, nutrient concentrations, and biological activity. These measurements are critical for understanding ocean circulation, stratification, and ecosystem health.
- **Monitor Ocean Currents and Climate Change**: Continuous monitoring of ocean currents, thermal structures, and chemical composition has provided insights into climate change, including the impact of global warming on ocean circulation patterns.

1.7.2 Marine Biology and Ecology

Marine biologists have leveraged UUVs to explore the rich biodiversity of the oceans for various areas of study.

- **Habitat Mapping**: AUVs equipped with video cameras, sonar, and environmental sensors are used to map coral reefs, seagrass beds, and deep-sea habitats (Figure 1.7).

FIGURE 1.7 C-Bot AUV developed by NIO Goa for monitoring coral reefs.

Source: **Dr Pramod Maurya.**[1]

- **Species Monitoring**: Autonomous vehicles enable the study of migratory patterns, behaviour, and population dynamics of marine species without disturbing their natural habitats. UUVs can be programmed to follow marine life over long distances, providing valuable data on species behaviour and ecosystem interactions.
- **Invasive Species and Biodiversity Assessments**: UUVs help monitor the spread of invasive species and assess the overall health and diversity of marine ecosystems, contributing to conservation efforts.

1.7.3 Underwater Archaeology

UUVs have opened new frontiers in underwater archaeology by enabling the exploration of shipwrecks, submerged cities, and ancient harbours.

- **High-Resolution Imaging**: AUVs with advanced imaging systems capture detailed visual and sonar data, revealing structural details of submerged archaeological sites.
- **Non-invasive Exploration**: Autonomous systems minimise human intervention, reducing the risk of damage to fragile historical sites.

In the 2010s, UUVs were deployed in the Mediterranean to map ancient shipwrecks and submerged settlements, yielding high-resolution 3D reconstructions that have been invaluable for historians and archaeologists. ROVs are being used by NIO Goa for the survey of underwater artefacts off Dwarka in Gujarat, India (Figure 1.8).

FIGURE 1.8 Underwater stone anchor off Dwarka as seen from an ROV.

Source: **CSIR-NIO Goa.**

1.7.4 Environmental Monitoring

Industrial activities, climate change, and urbanisation have all led to increased environmental stress on marine ecosystems. UUVs are now crucial for:

- **Pollution Detection**: Autonomous vehicles are used to monitor water quality, detect oil spills, and track chemical contaminants in the marine environment.
- **Microplastics Studies**: UUVs equipped with specialised sensors help researchers quantify and analyse the distribution of microplastics in different ocean regions.
- **Long-Term Monitoring**: Many research programmes deploy UUVs for long-term environmental monitoring, collecting continuous data that informs climate models and environmental policy.

1.7.5 Deep-Sea and Polar Research

The challenging conditions of deep-sea and polar regions require robust and highly autonomous vehicles. Vehicles such as the REMUS 6000 and HUGIN AUV have been used to explore deep-sea ecosystems, hydrothermal vent fields, and seafloor geology. These missions have provided new insights into extreme life forms and the dynamics of deep ocean processes.

In polar regions, UUVs are essential for studying ice-covered seas, monitoring ice melt, and understanding polar climate systems. Their ability to operate autonomously under extreme conditions is critical for gathering data where human access is limited (Butler, 2018).

1.8 INDUSTRIAL USES OF UUVS

Beyond scientific exploration, UUVs have found a wide range of industrial applications, particularly in sectors that require robust underwater inspection, monitoring, and maintenance (Stump, 2019).

1.8.1 Oil and Gas Industry

The offshore oil and gas sector has been an early adopter and promoter of UUV technology for various functions:

- **Pipeline and Cable Inspection**: UUVs are deployed to inspect underwater pipelines and cables for corrosion, damage, or displacement (Zhang, 2020). High-resolution sonar, video, and other sensors are used to detect issues that could lead to leaks or system failures (Figure 1.9).
- **Wellhead Monitoring**: Autonomous vehicles are used to monitor underwater wellheads, ensuring safe operations and compliance with environmental standards.

FIGURE 1.9 Underwater pipeline inspection in progress, monitored by ROV.

Source: **TSC Subsea.**[2]

- **Seabed Surveys**: Before drilling or pipeline installation, detailed seabed mapping is performed using UUVs. This data is critical for planning safe and efficient infrastructure projects.

In the 2000s, companies such as Oceaneering and Fugro began using UUVs extensively for pipeline inspection and subsea facility monitoring in the Gulf of Mexico and the North Sea.

1.8.2 Underwater Infrastructure Maintenance

UUVs play a crucial role in the construction and maintenance of underwater infrastructure, as discussed below.

- **Structural Inspections**: Unmanned vehicles are used to inspect offshore platforms, wind turbine foundations, and other marine (or any underwater) structures. These are highly useful for structural inspections of bridges and dams on inland water bodies as well. Their ability to provide close-up, high-definition images reduces the need for expensive and hazardous diver operations.
- **Repair and Maintenance Support**: Autonomous vehicles can carry tools and sensors to assess structural integrity, detect corrosion, and support maintenance operations. For structures such as bridges and dams, such surveys are part of mandatory safety requirements (Figure 1.10).

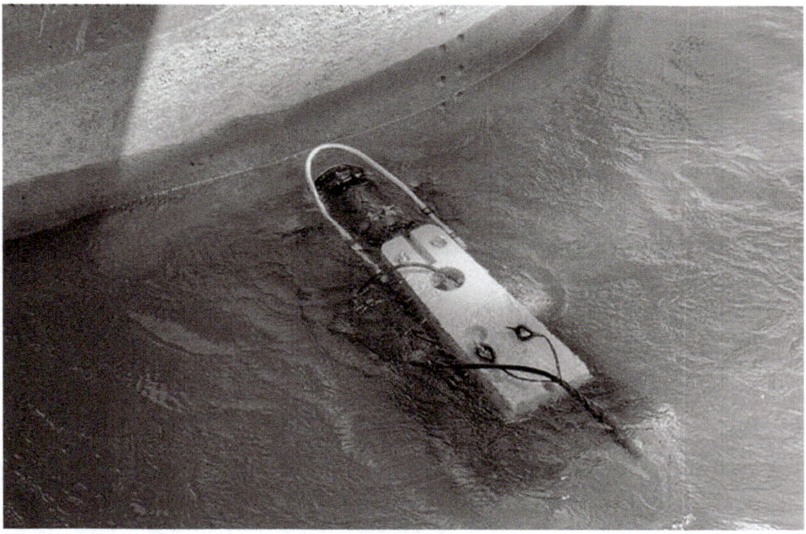

FIGURE 1.10 ROV Mikros inspecting bridge supports in India.

Source: **Planys Technologies (Setia, 2020).**

- **Port and Harbour Inspections**: UUVs conduct routine inspections of harbour facilities, including breakwaters, piers, and underwater lighting systems, to ensure structural integrity and safety.
- **Subsea Cable Monitoring**: As global data transmission increasingly relies on submarine cables, UUVs are deployed to inspect these cables for damage or tampering (Xiang Liu, 2025).
- **Underwater facilities**: Underwater housing of data centres and possibly offshore energy infrastructure in the future would require the use of UUVs for routine access, inspection and first line of maintenance.
- **Diver Safety**: By performing routine inspections and initial surveys, UUVs reduce the risks to human divers, especially in hazardous or deep-water environments (Pantheeradiyil, 2024).

Several offshore projects have employed UUVs for routine inspection and maintenance of offshore wind farms, significantly reducing operational downtime and increasing safety. UUVs are used for monitoring underwater cables to ensure uninterrupted communication services (Rumson, 2021).

1.8.3 Seabed Mining and Resource Exploration

As interest in marine mineral resources grows, UUVs are increasingly being used in the exploration and eventual extraction of underwater resources (Figure 1.11).

FIGURE 1.11 Deep-sea mining machine 'Varaha-I' of NIOT weighing 9.5 tonnes being lowered for operations in the Indian Ocean up to 5,300 metres depth.

Source: **NIOT (Shankar, 2022).**

- **Polymetallic Nodule Exploration**: UUVs equipped with high-resolution mapping and geophysical sensors are deployed to survey the seafloor for polymetallic nodules, which contain valuable metals such as manganese, cobalt, and nickel (Shankar, 2022).
- **Resource Mapping**: Detailed geophysical surveys by UUVs help in assessing the viability of seabed mining projects by providing critical data on resource distribution, sediment composition, and seabed stability.

1.8.4 Other Industrial Applications

Other industries also benefit from the use of UUVs:

- **Environmental Impact Assessments**: UUVs are employed to assess the impact of industrial activities on marine environments, including sediment sampling and water quality measurements.
- **Marine Renewable Energy**: For offshore wind and tidal energy projects, UUVs help survey and monitor installation sites, reducing the need for costly manned surveys.
- **Fisheries Management**: UUVs contribute to sustainable fisheries management by monitoring fish populations, mapping breeding grounds, and assessing the health of marine ecosystems (Figure 1.12 shows a USV being used for fish feeding).

FIGURE 1.12 Unmanned surface vehicle used for dispersing fish feed in aquaculture farms in India.

Source: **i4Marine Technologies (Rosenoegger, 2025).**

1.9 DEFENCE AND SECURITY USES

Naval underwater systems have historically driven underwater technology development, including UUVs. The objective is to remove humans from dangerous areas and use uncrewed systems for routine surveillance. These features require greater sensing and autonomous decision-making capacity in vehicles.

The use of UUVs in Naval operations can reduce risk to manned personnel as well as enable tasks otherwise not possible. One or more UUVs operating with a standard submarine can significantly increase military capability at low cost while reducing risk to assets and personnel. Unmanned systems can act as extended sensors and means of intervention by submarines, and also as an extension of units deployed for the detection of enemy submarines.

Defence systems are demanding autonomous air and marine vehicles for networked multi-vehicle systems, including coordinated operations of several AUVs. Heterogeneous robot teams (water, land, air) are also an important area of application, particularly for continuous monitoring.

1.9.1 Intelligence, Surveillance, Reconnaissance

Intelligence-gathering, Surveillance, and Reconnaissance (ISR) is a major application area for naval UUVs. UUVs can be valuable for clandestine information gathering in areas close to enemy coastlines. UUVs can augment the capabilities of current naval platforms for ISR with minimal costs.

1.9.2 Mine Countermeasures (MCM)

Mine Countermeasures (MCM) operations are critical for searching and disposing of explosives that could have been laid in shallow approaches to harbours, in order to protect valuable ships from damage by exploding underwater mines. Mine-hunting robot vehicles are being deployed on ships. UUVs used for MCM missions can reduce the exposure of personnel and high-value assets to dangerous environments such as minefields. UUVs offer advantages over traditional methods in MCM, especially in shallow water littoral regions and remote environments. Distributed multi-AUV systems, or combinations of ASV, AUV, and ROVs working in coordination, have also been demonstrated for naval mine countermeasure missions (Thales, 2020).

1.9.3 Anti-Submarine Warfare (ASW)

Anti-Submarine Warfare (ASW) is a major naval activity involving searching for enemy submarines, either to seek and track them in areas of interest, or to defend ships against submarine attack. Typically, such efforts involve several warships, helicopters, and fixed-wing aircraft, over hundreds of nautical miles, and for days or weeks at a time. Such operations may be initiated by the initial detection of a submarine passing through a continuously monitored location, and thereafter deploying aircraft or ships to detect the submarine using sonars, and thereafter maintain contact. UUVs can support the conduct of ASW missions by serving as mobile (although low-speed) and silent sensor nodes. Such concepts have been demonstrated at sea. Heterogeneous underwater networks for ASW are also being developed.

1.9.4 Object Detection and Recovery

Object detection and identification is an important naval capability to search for and recover objects from the seabed. These objects may be practice weapons, spent munitions, or accident wreckage. UUVs are essential for such search and intervention roles. Surveillance, inspection, and defence of underwater assets also require ROVs and AUVs. These roles can overlap with scientific/commercial applications like debris removal and search and retrieval. ROVs are also used in the military for deep-sea rescue operations for manned submarines.

1.9.5 Harbour/Port Security

Security of Harbours and Ports is a continuous requirement. UUVs can assist in providing better monitoring of these vital infrastructures against the threat of underwater attack or sabotage.

1.9.6 Inspection of Facilities and Assets

Underwater structures or platforms on land include bridges, dams, and industrial facilities linked to water bodies. These require routine inspections for structural integrity, which can be conveniently undertaken using UUV solutions. Underwater hulls of ships require inspection, cleaning, and repairs in dry dock.

Some of these inspection activities can be undertaken by ROVs, thereby reducing the requirement for divers or dry docking of ships (Figure 1.13). Coastal infrastructure such as sea walls, marinas, and offshore defence facilities can be surveyed by UUVs. ROVs are also used in nuclear industry inspections.

1.9.7 Payload Delivery and Strike

Payload delivery is a potential mission for UUVs to undertake clandestinely. Cargo-carrying AUVs have been explored as a concept. UUVs can be used for attacking targets in harbours or seabed infrastructure, referred to as Influence Activities (IA) and Time Critical Strike (TCS). These potential offensive roles for UUVs require range, speed, and sophisticated capabilities (Dickinson, 2025).

FIGURE 1.13 ROVs for underwater inspection of dams, bridges, offshore structures, and ships.

Source: **Planys Technologies.**[3]

1.10 COLLABORATIVE RESEARCH

The deployment of unmanned underwater systems offers high cost and safety benefits across sectors. In scientific and industrial contexts, they reduce the need for expensive crewed operations and minimise the risks to human life. In defence, the ability to operate unmanned systems in high-risk environments provides a force-multiplying effect while preserving valuable personnel. However, balancing cost against performance remains a universal challenge, and investments in research and development are critical to achieving these efficiencies. Historically, most of the developments in oceanographic sciences in the 20th century can be linked to projects supported by the Navy (Weir, 2001).

Opportunities exist for collaborative research across domains, which should be leveraged for common applications.

- **Academic-Industry Partnerships**: Joint ventures between research institutions and industrial firms are driving innovations in sensor technologies, AI-driven autonomy, and energy storage. These partnerships can accelerate the transfer of cutting-edge technologies from scientific prototypes to industrial and defence applications.
- **Cross-Domain Demonstrations**: Pilot projects that test unmanned systems in multipurpose missions can highlight the adaptability of the technology. For instance, an AUV designed for scientific mapping might also be tested in industrial inspection scenarios, leading to design modifications that benefit both sectors.
- **Standardisation and Interoperability**: The establishment of industry standards for communication protocols, data formats, and interface specifications can promote interoperability between systems used in different domains, enhancing overall operational efficiency and reducing duplication of effort.

In the years to come, the marine robotics industry is projected to grow in value and reach (UUV Industry Research Report, 2024). Having seen an overview of the types of uncrewed underwater vehicles and their applications, the next chapter discusses the approaches for their design.

NOTES

1. CSIR-National Institute of Oceanography, Goa; *Ocean Digest*, Vol. 10, Issue 3, July 2023, Ocean Society of India.
2. https://www.tscsubsea.com/pipeline-inspections/.
3. https://planystech.com/rovs/.

REFERENCES

Ackerman, E. (2019, July 25). *Meet Aquanaut, the Underwater Transformer.* IEEE Spectrum. https://spectrum.ieee.org/meet-aquanaut-the-underwater-transformer

American Bureau of Shipping. (2025). *Rules for Building and Classing Underwater Vehicles, Systems and Hyperbaric Facilities.*

Aoki, T. (2008). Autonomous Underwater Vehicles: Design and Operation. In T. Ura (Ed.), *Urashima: Autonomous Underwater Vehicles, Design and Operation* (2nd ed.). Ura.

Ballard, R. D., & Drew, C. (2021). *Into the Deep: A Memoir from the Man who Found Titanic.* National Geographic.

Busby, R. (1976). *Manned Submersibles.* Office of the Oceanographer of the Navy.

Butler, B. (2018). *Into the Labyrinth: The Making of a Modern-Day Theseus.* Canada.

Choi, H.-T., & Yuh, J. (2016). Underwater Robots. In B. Siciliano & O. Khatib (Eds.), *Springer Handbooks* (pp. 595–622). Heidelberg: Springer International Publishing. https://doi.org/10.1007/978-3-319-32552-1_25

Christ, R., Wernli, R., & Wernli Sr, R. (2011). *The ROV Manual: A User Guide for Observation Class Remotely Operated Vehicles.* Netherlands: Elsevier Science.

Christ, R. S. (2011). *The ROV Manual: A User Guide for Observation Class Remotely Operated Vehicles.* Netherlands: Elsevier Science.

Davis, R., Eriksen, C., & Jones, C. (2002). Autonomous Buoyancy-Driven Underwater Gliders. In G. Griffiths (Ed.), *The Technology and Applications of Autonomous Underwater Vehicles.* London: Taylor and Francis. https://doi.org/10.1201/9780203522301

Dickinson, P. (2025, June 12). Ukraine Is Shaping the Future of Drone Warfare at Sea as Well as on Land. *AtlanticCouncil.org.* https://www.atlanticcouncil.org/blogs/ukrainealert/ukraine-is-shaping-the-future-of-drone-warfare-at-sea-as-well-as-on-land/

Geyer, R. (1977). *Submersibles and their Use in Oceanography and Ocean Engineering.* New York: Elsevier.

Griffiths, G. (2003). *Technology and Applications of Autonomous Underwater Vehicles.* (G. Griffiths, Ed.) Taylor & Francis. https://doi.org/10.1201/9780203023249

Moore, S., Bohm, H., & Jensen, V. (2010). *Underwater Robotics: Science, Design and Fabrication.* Monterey, CA: Marine Advanced Technology Education Center.

Pantheeradiyil, A. (2024, October 7). How ROVs and Divers Are Transforming Underwater Inspections for Safety and Efficiency. *blueyerobotics.com.* https://www.blueyerobotics.com/blog/how-rovs-and-divers-are-transforming-underwater-inspections-for-safety-and

Ray, A., Singh, S. N., & Seshadri, V. (2011, June 29–30). Underwater Gliders - Force Multipliers for Naval Roles. *Warship 2011: Naval Submarines and UUVs.* Royal Institution of Naval Architects.

Rosenoegger, J. (2025, May 15). India's i4 Marine Develops Smart Water Solutions for Sustainable Future. *Maritime Innovations.* https://maritime-innovations.com/i4-marine-technologies-water-innovation/

Rumson, A. G. (2021). The Application of Fully Unmanned Robotic Systems for Inspection of Subsea Pipelines. *Ocean Engineering, 235.* https://doi.org/10.1016/j.oceaneng.2021.109214.

Setia, R. (2020, December 3). Dinesh Natarajan, R&D Head of Planys Technologies on How the Company is Redefining Underwater Robotic Inspections with Remotely Operated Vehicles (ROVs). *Circuit Digest.* https://circuitdigest.com/interview/dinesh-natarajan-rnd-head-of-planys-technologies-on-how-the-company-is-redefining-the-underwater-robotic-inspections-with-remotely-operated-vehicles

Shankar, P. (2022, October 26). Deep-sea Mining Efforts Gear Up to Meet Clean Energy Demands Amid Concerns. *Mongabay.* https://india.mongabay.com/2022/10/deep-sea-mining-efforts-gear-up-to-meet-clean-energy-demands-amid-concerns/

Stump, J. (2019, March 2). Industry Advances ROV and AUV Technologies. Offshore. https://www.offshore-mag.com/subsea/article/16764019/industry-advances-rov-and-auv-technologies

Sweeney, J. B. (1970). *A Pictorial History of Oceanographic Submersibles.* London: Robert Hale & Company.

Thales to deliver the World's First Fully Integrated Unmanned Mine Countermeasures System for the Royal Navy and French "Marine Nationale. (2020, November 26). thalesgroup.com: https://www.thalesgroup.com/en/group/journalist/ press_release/thales-deliver-worlds-first-fully-integrated-unmanned-mine

Unmanned Underwater Vehicles (UUV) Industry Research Report 2023-2030: Opportunities in Deep-Sea Exploration, Subsea Resource Mapping, Maritime Surveillance and Defense Applications. (2024, September 03). https://www.globenewswire.com/: https://www.globenewswire.com/news-release/2024/09/03/2939731/28124/en/Unmanned-Underwater-Vehicles-UUV-Industry-Research-Report-2023-2030-Opportunities-in-Deep-Sea-Exploration-Subsea-Resource-Mapping-Maritime-Surveillance-and-Defense-Applications.html

Weir, G. E. (2001). *An Ocean in Common: American Naval Officers, Scientists, and the Ocean Environment.* Texas: A&M University Press.

Xiang Liu, S. X. (2025). Innovative Strategy and Practice of Using Underwater Robot for Marine Cable Inspection and Operation and Maintenance. *Cognitive Robotics.* https://doi.org/10.1016/j.rineng.2021.100201

Yoerger, D. R., Bradley, A. M., Jakuba, M., German, C. R., Shank, T., & Tivey, M. (2007, March). Autonomous and Remotely Operated Vehicle Technology for Hydrothermal Vent Discovery, Exploration and Sampling. *Oceanography*, *20*(1), 152–161. Retrieved September 2025, from https://oceanrep.geomar.de/id/eprint/38822/1/Yoerger.pdf

Zhang, H., et al. (2020). Subsea Pipeline Leak Inspection by Autonomous Underwater Vehicle. *Applied Ocean Research.* https://doi.org/10.1016/j.apor.2020.102321

2 Design Considerations and Standards

2.1 APPROACHING THE TASK

Designing a product starts with identifying and analysing the requirements, examining alternative concepts to satisfy those requirements, followed by the definition of components and analysis of performance. Iterations would be required as changes in one aspect or component of the design would impact its other features. The physical building blocks would comprise components that are either already developed or under development, and this level of maturity would affect the level of uncertainty or the number of iterations required in the design.

Design of an AUV or ROV is more challenging than most other engineering creations due to the extreme environmental conditions in the depths of the seas. While UUV design is not governed by the degree of risk that pervades human-occupied submersibles, UUVs do need to operate autonomously and reliably, even without communications (Busby, 1976).

Any design is governed by a trade-off between critical mission requirements and constraints on fabrication, assembly, and operational logistics (Ullman, 2017). This process for a UUV involves significant effort and ingenious engineering solutions. The design methodology can be viewed as a sequence of stages and procedures to plan, define, analyse, describe, build, test, and induct a specific product or solution (Dieter & Schmidt, 2021).

2.1.1 Design Methodology

A generic methodology for designing underwater vehicles should include the following stages:

- **Understand the Mission**: Examine the range of operational requirements (or gaps in capability from market surveys) and select the key mission that your UUV should aim to achieve (Geyer, 1977). This should be distilled into clear objectives (a few statements), minimising labels or jargon.
- **Develop the Mission Tasks**: Break down the mission statement into a step-by-step series of specific actions the vehicle must be able to perform. In naval parlance, this would be called 'Concept of Operations' or CONOPS.
- **Establish Performance Requirements**: Define how well the system must perform the identified tasks. This would include operational parameters to

DOI: 10.1201/9781003737476-2

be achieved by the eventual design, such as speed, range, endurance, depth, or specific task performance metrics.

- **Identify Constraints**: Recognising limitations such as time, budget, access to tools and facilities, regulations, and environmental conditions. Design parameters are often influenced more by constraints than by a 'wish list' of performance metrics.
- **Select Sensors/Technologies/Systems**: Identify the major sensors and subsystems that are needed to perform the mission tasks. Select the major technology solutions to be adopted, depending on performance requirements, constraints, and risk/maturity of each technology solution. This stage could require extensive data collection, but it needs to be a time-bound activity, and this is where experience and a mature industrial base make a difference.
- **Generate Alternative Concepts**: Develop one (or more) concept design configurations that integrate the identified systems, considering technical feasibility within constraints. This stage marks the transition of project objectives into a viable design solution. Every concept or solution represents not only combinations but also compromises. Generating more than one solution would enable a wider exploration of alternatives, either in terms of varying performance requirements or technology/component choices.
- **Preliminary Design**: For the selected concept design of the UUV, analysis of each of its component systems is undertaken to verify the assumptions in the previous stage, based on more concrete data and using more elaborate tools. Results of analysis of one aspect may require changes in other aspects, which could lead to repeating the analysis for those affected aspects as well (Jackson, 1992). This is the iterative nature of design, but it needs to be capped by thoughtful choices, mature data, and exercise of judgement by the head designer (taking hard decisions!).
- **Detailed Design**: The design has to be broken down into the definition of each of its systems or components for each manufacturer or supplier (or integrator) to comply. This is more time-consuming and requires great attention to detail. Iterations are not welcome at this stage. The completeness of manufacturing/integration details, consistency of information, and correctness of interfaces are the focus areas in this stage.
- **Construction**: Building a physical prototype of the UUV needs to be as close to the intended production version as feasible. Physical prototypes can be as expensive or complex as the eventual product, and there is limited scope for experimentation in major choices. Ideally, new technologies should have been de-risked separately through selective installation or evaluation on other test platforms. The complete UUV prototype should aim for minimum surprises and should focus on demonstration of capability (rather than a test bed for the designer).
- **Trials and Evaluation**: After the prototype is functionally satisfactory, its performance characteristics need to be demonstrated in trials that need to be defined for acceptance of the final product. Extensive trials of the prototype would offer insights into aspects that may not have been explicitly

specified as requirements, but still need refinement. Even after a design has entered the production stage, the designer needs to remain engaged with operational feedback from the user. This would provide essential feedback to identify areas of technology development, pain points that need to be addressed in future design, and unforeseen ways of utilisation that could suggest new designs.

2.1.2 Design Quality

AUV design involves multidisciplinary optimisation techniques, aggregating individual subsystem analyses using system-level coordination. Designing for the specific mission and environment is crucial, but being able to utilise the designed product for a wider range of missions (possibly with minor upgrades or modifications) indicates the robustness of the design.

Introduction of breakthrough technologies cannot be expected to transition immediately into widely adopted products. Reliable performance parameters and ease of maintenance and operation would be the defining characteristics of a great design.

The designer must remember that any UUV is only the means to an end. That end objective should be well understood (even if not explicitly articulated) by the designer, which requires awareness and appreciation of operating conditions at sea, facilities and constraints for deployment, and the level of training of operators.

It is important to remember that every 'uncrewed' system does require a human crew to deploy, operate, and maintain, both at sea and ashore. User experience and satisfaction are therefore as essential in the design of UUVs as in 'manned' submersibles.

2.2 TYPICAL PAYLOAD AND LAYOUT

The term 'payload' refers to equipment that meets the primary purpose for which the vessel is deployed. For a UUV intended for survey, the payload would consist of cameras or side scan sonar. For an ROV intended for object recovery, the payload would be the manipulator. In general, payload would include those equipment of the UUV that are essential for undertaking operations, as well as sensing its environment and collecting environmental data. Every vessel (UUV or ROV) must be designed around its mission or purpose, which translates to selection of payload first, and then designing the vehicle around the main elements of payload. Examples of two Indian AUVs and their layout are shown in Figures 2.1 and 2.2.

2.3 SYSTEMS TO BE DESIGNED

The main systems and components of an AUV or ROV include physical components, software (crucial for autonomous operation), and support systems ashore or on the mother ship.

The major aspects of design of autonomous underwater vehicles include structural design, hydrodynamics, powering, energy storage, sensors, controls, and

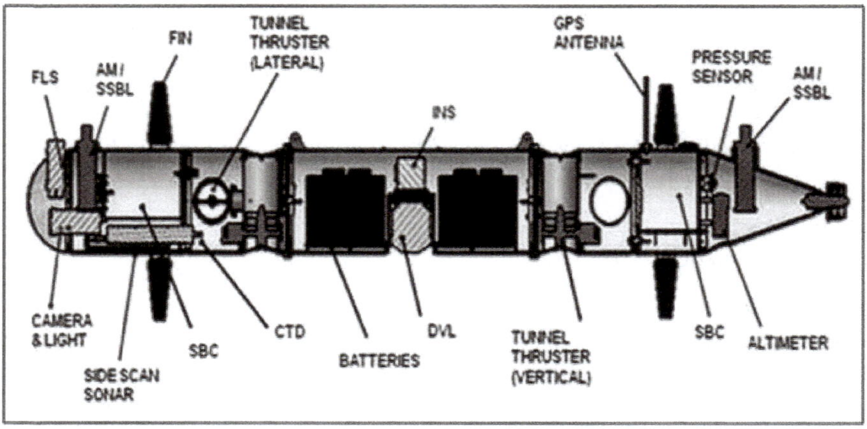

FIGURE 2.1 Components and Layout of AUV-150 developed by CSIR-CMERI Durgapur, India.

Note: **Ambastha et al., 2014.**

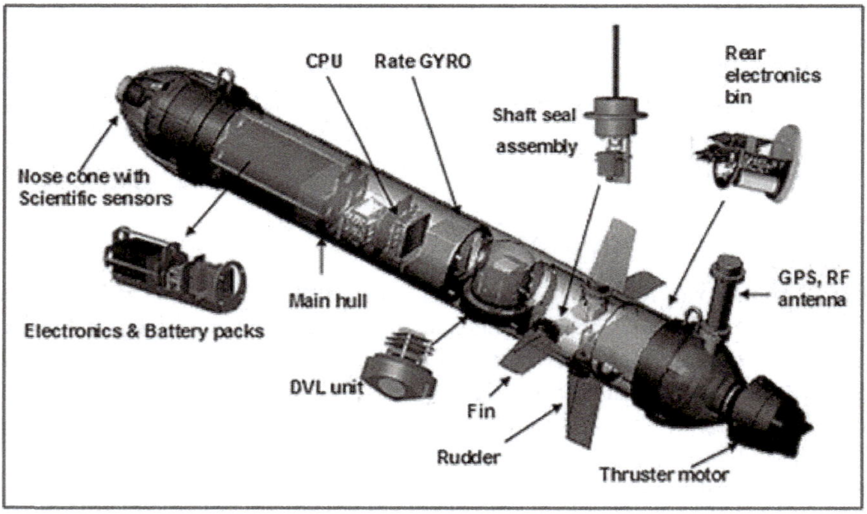

FIGURE 2.2 Components and layout of AUV Maya developed by CSIR-NIO Goa.

Note: **Madhan et al., 2007.**

computational modules. The subsequent sections of this chapter discuss these design aspects in qualitative terms. These will be covered in further detail in subsequent respective chapters.

2.3.1 Hydrodynamics and Weight

The hydrodynamic characteristics significantly impact an AUV's performance. Hull shape affects drag, speed, and manoeuvrability. Structural strength is essential to withstand hydrostatic pressure at depth.

- **Hull Forms**: Torpedo-shaped bodies are common for cruising vehicles (AUVs) needing low drag at velocity, often using a main thruster and control surfaces. Box-like shapes or cages with easy access to components and multiple thrusters are used for vehicles working mostly at low velocity (such as ROVs). Spherical shapes have also been designed. Hybrid glider concepts exploit gravity-buoyancy forces for slow, long-duration movement, including wing-like shapes or axisymmetric cylinders with appended wings. Biomimetic designs mimic fish for manoeuvrability and in the quest for greater energy efficiency.
- **Weight**: Any underwater vehicle must match its weight with buoyancy. Weight control and means for adjusting weight/buoyancy are therefore fundamental. Materials must be corrosion-resistant for the seawater environment. The deeper the depth rating, the heavier the buoyant structure (pressure-resistant enclosure or 'pressure hull') must be to withstand hydrostatic pressure; hence, materials with a greater strength-to-weight ratio are preferred.
- **Buoyancy and Ballast**: The ballast system adjusts the vehicle's tendency to drift upwards or downwards. AUVs generally need to maintain near-neutral buoyancy. Ballast systems can range from simple static weights or foam to complex variable buoyancy systems using pumps, pistons, or compressed gas.

2.3.2 Structural Design

The size of the equipment that needs to be protected from exposure to seawater is the main consideration for deciding the diameter and length of the pressure-resistant section(s) of the AUV.

The structural design of the pressure-resistant portion is based on the load due to external hydrostatic pressure at the maximum operating depth for which the AUV is to be designed.

This structural load is usually so high that any other loads are rendered relatively insignificant for the design of the pressure hull. The shape selected is either a sphere, which is difficult to fabricate, or a cylinder with hemispherical (or 'dished') ends, which offers greater usable internal space and is more conducive to modular arrangement. The cylinder is usually stiffened by circular ring frames (longitudinal stiffening is not necessary).

TABLE 2.1
High-Strength Materials for Underwater Structural Applications

Material	Specific Density (g/cm³)	Young's Modulus (GPa)	Compressive Yield Strength (MPa)	Ultimate Strength (MPa)	Poisson Ratio
Aluminium (6061)	2.700	70	276	310	0.33
Stainless steel (316)	8.000	195	290	550	0.30
HY80 steel	7.828	210	550	690	0.29
HY100 steel	7.828	210	690	793.5	0.29
Titanium alloy (Ti-6Al-4V)	4.430	110	827	1000	0.30
ABS (moulded)	1.050	2.2	65	40	0.35
Polycarbonate (extruded)	1.200	2.4	70	67	0.37

Notes: Compiled from Allmendinger, 1990; Fathallah & Helal, April 2016; Moore, Bohm, & Jensen, 2010

2.3.3 Structural Materials

Materials used for various UUV components include fibre composites, ceramics, syntactic foam, aluminium, steel, high-density polyethylene, and glass-reinforced plastics. Practicality, manufacturability, and ease of maintenance are important factors for the material selection of components.

Pressure hulls are critical components, often designed as ring-stiffened cylinders or unstiffened monocoques. The material of the pressure hull structure needs to have a high strength-to-weight ratio, display a suitable degree of elasticity, withstand corrosive environments and cyclic (fatigue) loading, and be feasible to fabricate. Such metals include steel, aluminium or titanium alloy. Composite materials such as GRP or CFRP suffer from the disadvantages of nonlinear behaviour under loading, and difficulty in inspection or repairs. Typical examples are given in Table 2.1.

Other materials commonly used for structural components in underwater vehicles are listed in Table 2.2.

2.3.4 Propulsion and Actuation

Propulsion systems generate the forces needed for the movement of the UUV. They can include thrusters and control surfaces (rudders, elevators).

- **Thruster Types**: The main propulsion mechanism for the majority of AUVs is an axially mounted propeller, usually referred to as 'thruster' in combination with the motor that turns the propeller for axial thrust. Fixed propulsion uses actuators (which could be thrusters) to generate force in a fixed direction relative to the body. Vectorial thrusters can change the direction

TABLE 2.2
Commonly Used Structural Materials in Underwater Vehicles

Material Type	Material	Used in Frames	Used in Pressure Hull
Metals	Steels	Yes	Yes
	Aluminium	Yes	Yes
	Titanium	Yes	Yes
Plastics	Acrylic	No	Yes (viewports)
	ABS, PVC	Yes (brackets, parts)	Yes (shallow water)
Composites	GRP	Yes (fairings)	Experimental
	Syntactic foam	Floatation	No
Other	Glass	No	Experimental
	Ceramic	No	Experimental
	Rubber	Bumpers	O-Ring Seals
	Lead	Yes (ballast)	No

Notes: Adapted from Moore, Bohm, & Jensen, 2010

of force. The number, orientation, and position of thrusters depend on the mission.

- **Degree of Actuation**: The degrees of freedom (DOFs) that are actuated or actively controlled are a measure of the vehicle's manoeuvrability, especially at low forward speed. Fully actuated UUVs can control all six DOFs of their motion, while under-actuated vehicles control fewer (maybe only one forward DOF and two angular DOFs). Fully actuated vehicles are good test beds for control methods due to high manoeuvrability and are more significant for ROVs, while industrial AUVs often prioritise autonomy and velocity for survey tasks, where full actuation is less critical.
- **Control Surfaces**: These refer to the rudders and planes, often used in conjunction with thrusters, especially for cruising vehicles. Their deflection (degree of turning) provides lift and directional control.
- **Control Allocation**: For vehicles with multiple actuators (over-actuated), control allocation distributes desired control forces across actuators. The objective is often to minimise energy use while satisfying constraints. Techniques include generalised inverse or quadratic programming.

2.3.5 Energy Source

Energy capacity is an important constraint of a UUV, limiting the cruising range of AUVs and the mode of operating a mission.

- **Batteries**: Common power sources include lithium-ion, lithium-polymer, nickel-metal hydride, and lead-acid batteries. Lithium-ion and lithium-polymer offer higher energy density and longer cycle life. Battery technology is a key constraint on mission duration. Batteries can account for 5% to 45% of an AUV's mass.

- **Fuel Cells**: Proton exchange membrane fuel cells (PEMFC) are suitable for underwater applications. They offer high energy density for the generator itself, but the system's weight increases due to tanks and auxiliary components. They hold promise for long endurance missions.
- **Buoyancy Engines**: Underwater Gliders are designed for very long endurance, actively changing their buoyancy to take advantage of their shape and utilise (passive) downward/upward motion for achieving forward movement as well. The actuation of the buoyancy engines also requires (much lesser) power from on-board batteries. Hybrid gliders can have a combination of thrusters and buoyancy engines for greater mission flexibility.

2.3.6 Navigation and Positioning Systems

Accurate navigation and positioning are fundamental for any mission underwater. GPS can be used only intermittently, when the UUV is on the surface. While underwater, the UUV is dependent on a combination of inertial navigation, supplemented by inputs from other sensors.

- **Systems**: On-board systems often include inertial navigation systems (INS), Doppler Velocity Logs (DVL), depth sensors, and compasses. GPS can be used only when the vehicle is on the sea surface, with the antenna exposed.
- **Acoustic Positioning**: Underwater positioning relies on acoustic systems. Super-short baseline (SSBL), also called ultra-short baseline (USBL), systems track the AUV from a surface vessel. Long baseline (LBL) systems use seafloor transponders.
- **Sensor Fusion**: Data from multiple navigation sensors is often merged and correlated to improve accuracy and manage error growth. Techniques like Kalman filters (EKF, AFKF) and fuzzy logic are applied.
- **Challenges**: Navigation is challenging underwater due to a lack of GPS, signal attenuation, and dynamic currents. Position estimation error can drift without external resets. Local (relative) navigation is generally more accurate than global navigation.

2.3.7 Sensor Payloads

A comprehensive sensor suite is essential for a UUV to operate underwater, for external data collection and reliable navigation. This includes the following:

- **Inertial Measurement Units (IMU)**: Provide acceleration and angular rate data for motion estimation.
- **Doppler Velocity Log (DVL)**: Measures the vehicle's velocity relative to the seabed.
- **Pressure Sensors**: Measure depth and help with buoyancy control.

- **Sonar and Acoustic Sensors**: Facilitate obstacle detection, mapping, and communication.
- **Optical Sensors and Cameras**: Used for visual navigation, object recognition, and environmental monitoring.
- **Environmental Sensors**: Temperature, salinity, and current sensors help the vehicle adapt to changing conditions.

Some of the sensors are themselves examples of 'payload' in the sense that they carry out observations not only to support the operations of the UUV but also collect information that forms the mission of the UUV. Thus, the payload could include environmental sensors for a scientific AUV, the optical sensors for a scientific or industrial AUV, and the acoustic sensors for a military AUV.

The choice of sensor suite on a particular AUV design is oriented towards its mission, so that the requirements of 'payload' can be combined to the extent feasible with the instruments required for the navigation and operation of the vehicle itself.

Beyond navigation, AUVs carry sensors for mission-specific data collection. These are essential for the core mission of the UUV, and the vehicle must be designed around its payload.

- **Types**: Common mission sensors include sonar (side-scan, multibeam echo sounder, sub-bottom profiler, obstacle avoidance sonar, forward-looking sonar, DIDSON, magnetic tracker), cameras (digital, TV, CCD), CTD, magnetometers. Specific sensors are developed for biological, chemical, geological, and physical oceanography.
- **Integration**: Sensors must be integrated mechanically, electronically, and via software. Packaging in pressure vessels is common. Sensor size, weight, power consumption, data storage, and bandwidth requirements must be considered during design. Designing sensors compatible with multiple AUV types adds complexity. Sensors should ideally have minimal energy requirements and should not significantly impact the AUV's hydrodynamics, but their operational requirements should be given primacy in the design.

2.3.8 Communication Systems

Underwater communication has a limited range and bandwidth, presenting a major challenge.

Acoustic communication is the primary method underwater. Radio frequency (RF) and satellite links require surfacing or tethered modules. Optical links are still at experimental stages.

Designing effective communication networks for multiple AUVs or AUVs interacting with surface/other assets is complex. Multi-hop communication techniques can extend the range. Network topology (centralised vs. decentralised) impacts communication needs.

2.3.9 Control System

Controlling the trajectory (path) and behaviour of AUVs is challenging due to nonlinear and coupled dynamics, difficulty in estimating parameters, and environmental disturbances. The control system of an AUV manages sensor and actuator systems.

The design of any control system aims to ensure accurate manoeuvring, trajectory tracking, station keeping, robust behaviour under disturbances and uncertainties, while minimising power consumption.

Common methods or approaches for control include the following: Proportional-Integral-Derivative (PID), Sliding Mode Control (SMC), Adaptive Control (AC), Fuzzy Logic, Model Predictive Control (MPC), Computed Torque Control, and Backstepping Integral Adaptive (BIA). Nonlinear control methods are often needed for complex dynamics. Adaptive strategies or robust approaches are used to handle model uncertainty.

2.3.10 Autonomous System and Mission Computer

The autonomous system of an AUV provides the 'intelligence' for independent operation. Some basic concepts are outlined here:

- **Levels of Autonomy**: Autonomy is not binary but exists in levels, from teleoperation (remote human control) to fully intelligent systems. The level of autonomy is a critical design choice that impacts mission complexity and vehicle capability. Higher levels enable more complex missions.
- **Control Architectures**: Software frameworks manage sensors and actuators to execute missions (Autosar, n.d.). Approaches include Deliberate (hierarchical, Sense-Plan-Act), Reactive (based on behaviours, rapid response), and Hybrid architectures combining both. Hybrid systems plan at higher layers while lower layers handle adjustments and anti-collision (Hwang et al., 2014).
- **Planning and Decision-Making**: AUVs need on-board intelligence to react to unexpected situations. Mission planning defines waypoints and parameters, while dynamic mission planning or re-planning adapts the plan based on real-time information (environment, obstacles, target status). Path planning generates trajectories between points or covers areas. Coverage path planning is important for survey missions. Decision-making involves selecting advantageous actions based on available information. Artificial intelligence, neural networks, fuzzy logic, and knowledge-based systems are used for high-level control and decision-making.
- **Fault Detection and Tolerance**: The ability to handle internal malfunctions (sustainability) and unpredictable external events (adaptivity) is essential for achieving true autonomy. Fault detection/tolerant schemes involve monitoring vehicle condition and adapting to recover from faults or failures. Knowledge bases and reasoning can support fault tolerance.

- **Agent-Based Systems**: 'Agents' are specialised in disciplines and provide capabilities as services. Architectures using embedded agents are being developed, often based on service-oriented computing. This allows for modular integration and collaborative missions. The Operator Control Unit (OCU) interacts with agents in the vehicles. Planning can be centralised (at the OCU) or distributed (among agents).

The autonomous system or 'autonomy stack' can also be viewed in terms of conceptual 'layers', such as the physical layer (sensors, actuators and computational hardware), the interfaces layer, and the processing layer (Moos-Ivp, n.d., Ros, n.d.). The software architecture of AUVs forms the core of its autonomous capability and is discussed further in Chapter 8.

A service-oriented approach allows vehicle modules wrapped in services to be advertised and discovered (Agha, 1990). Knowledge bases capture domain expertise and dynamic information to support reasoning agents.

2.3.11 Data Processing and Management

AUVs collect significant quantities of data. Data is often stored on-board until recovery due to communication limitations. Future vehicles would have increasing requirements for on-board processing (such as for target recognition and analysis). Processing requires power and computational resources.

- **Processing Hardware**: On-board processing is essential for control, navigation, and communication. A distributed CPU system reduces processing load and allows redundancy. The Controller Area Network (CAN) bus is commonly used as an internal bus.
- **Software Architecture**: Software architecture design principles include organising decision-making to be decentralised, or hierarchical. Software modules are often distributed across on-board computers. Modularity allows easy addition, removal, and replacement of components. AUV software architecture is elaborated further in Chapter 8.
- **Data Integration and Fusion**: Data from various sensors on the UUV needs to be integrated. Sensor fusion combines data to locate the vehicle and targets. Data from integral sensors of one UUV may also need to be combined with data from other sources for overall situational awareness. Knowledge bases and the design of appropriate ontology support the representation and processing of information.
- **Data Transfer**: Mechanisms are needed for transferring data to ashore systems. High-speed data transfer methods are desired for sensor data, which need to be optimised within physical limitations, power requirements, and availability of bandwidth.

2.4 INITIAL SIZING

The traditional approach for ship or submarine design starts with the formulation of requirements, followed by progressive estimation of main parameters and performance, with iterations to develop each of the subsystems in further detail. We have outlined the steps of design in the first section of this chapter. But how does one estimate the initial size or form or configuration of a vehicle from abstract or flexible requirements? Some typical approaches, adapted for submarine or submersible design practices, are outlined in the following subsections.

2.4.1 'Type Ship' or 'Parent-Ship' Approach

The parent-ship approach involves scaling of an existing design to meet modified requirements of a new design. This method works best when complete details of the parent vehicle are available, and the changes envisaged are slight (say of the order of 10%), preferably for a single major feature or performance parameter.

However, changes are usually expected in more than one parameter of an existing vehicle design, and each parameter may have different correlations for scaling, which need to be physically meaningful and consistent. For example, the power requirement is proportional to the cube of speed. Internal volume, affecting fuel/energy content, is proportional to the cube of linear dimension. However, change in form would affect drag, power consumption, and endurance in ways that need to be considered in terms of physical effects on skin friction drag as well as viscous pressure drag, and may not be amenable to scaling by simple ratios.

2.4.2 'Building Block' Approach

This approach starts with the general arrangement of the major components within the overall envelope of the vehicle.

Most of the internal volume of AUVs is devoted to energy storage, typically in the form of batteries. 'Payload' refers to the instruments, sensors, or other objects that form the main purpose of deploying the AUV. The other major components are the thruster (usually an integrated motor with propeller) and the electronics. The electronics include power distribution, mission computers, data storage media, data buses, etc. Typical 'blocks' of the internal volume of an AUV are depicted in Figure 2.3.

The electronics are invariably located in a pressure-proof enclosure, which is cylindrical in shape for optimal strength against external hydrostatic pressure (Figure 2.4). Items located outside the pressure-proof enclosure (called 'pressure hull' on submarines) are electrically connected to the internal electronics using pressure-resistant cables. All cables passing through the pressure hull require special 'penetrators' at the boundary to prevent ingress of seawater, and therefore need to be minimised in quantity.

Based on the requirements of payload, endurance, speed and other performance parameters, the initial estimate of internal volume can be made from the volume

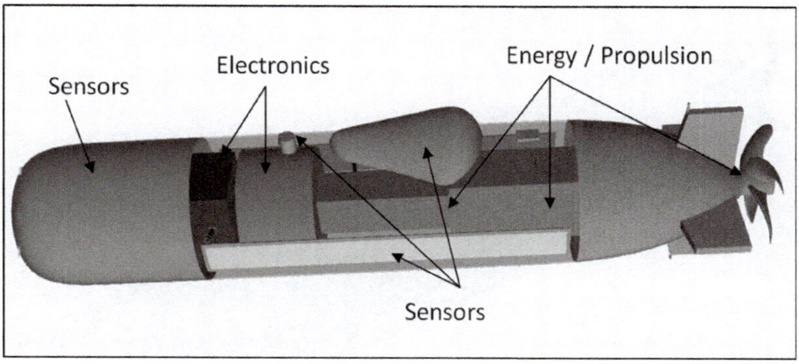

FIGURE 2.3 Blocks for equipment and systems in a typical AUV.

Source: **Author.**

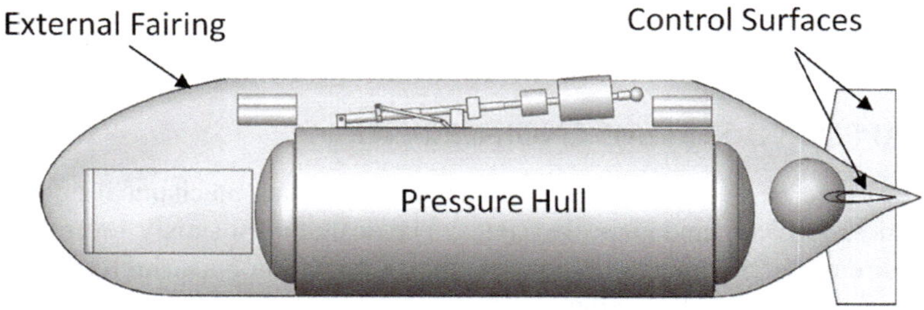

FIGURE 2.4 Pressure-proof and non-pressure-proof fairing.

Source: **Author.**

required for the major building blocks. These can then be refined considering a range of length and diameter values to conduct a parametric survey for correlating internal volume, size, form, and speed/endurance.

2.4.3 Equipment-Driven Design

For small vehicles such as AUVs, the selection of one or a few major equipment items such as a type of battery, or multibeam echosounder, or side scan sonar, could become a major factor affecting their size and form (Figure 2.5). These may also form the costliest as well as most significant components of the AUV (in terms of vehicle role or purpose). It would be prudent to start the design by identifying such major equipment and then design a cylindrical (or similar axisymmetric) enclosure to house them, as required for their functioning. From the viewpoint of functionality, this approach may result in a better estimate of vehicle weight, volume and form.

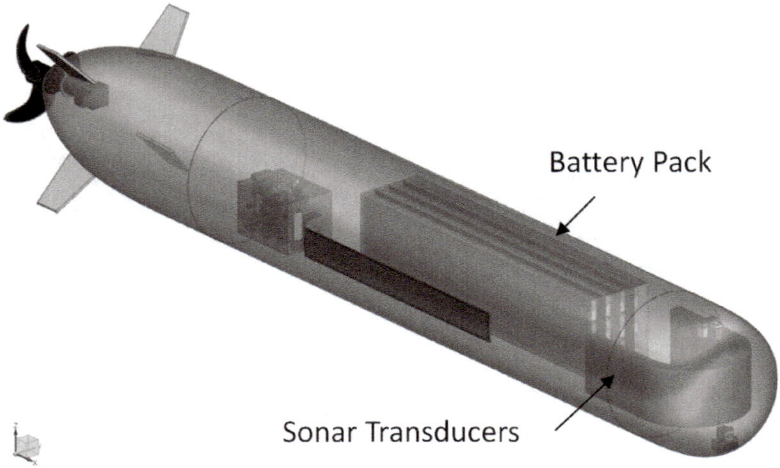

FIGURE 2.5 Examples of major equipment that may drive AUV design.

Source: **Author.**

2.5 APPLYING SYSTEMS ENGINEERING

AUV design is a multidisciplinary endeavour that spans mechanical engineering, electronics, software, and control systems. The design must satisfy several conflicting requirements: high energy efficiency, robust operation under high pressure, low maintenance, and high autonomy.

The 'Systems Engineering' (SE) approach, first formalised for aerospace applications, is a structured framework for designing, developing, and integrating complex systems (NASA, 2017). It provides a standardised methodology of progressing from user requirements to analysis and testing, with documentation and reviews at various stages. It encourages a 'systems' view of any design task, starting with a deep understanding of user/mission requirements. This section discusses key SE methodologies applied to AUV development.

2.5.1 Vee Model of SE

The 'Vee' model is a widely used representation in systems engineering to explain the decomposition of requirements, followed by the integration and verification of subsystems. The left arm of the 'V' (Figure 2.6) represents (moving downward) the progressive decomposition of system requirements and the design process, while the right arm (moving upward) represents system integration and validation.

- **Requirements Definition**: At the apex of the left side, stakeholders define mission requirements, operational constraints, and performance objectives.
- **Functional Decomposition**: These requirements are then decomposed into functional and technical specifications that guide the design of both hardware and software subsystems.

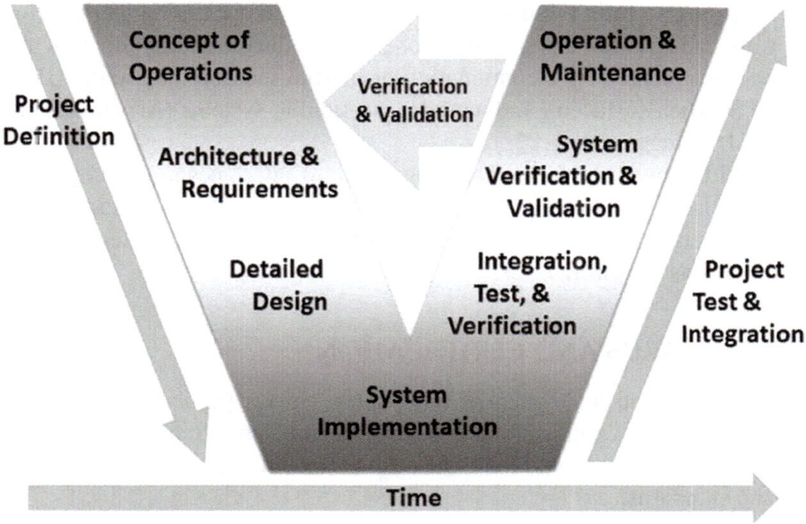

FIGURE 2.6 Graphical 'Vee' model in systems engineering.

Source: INCOSE, 2015.

- **Design and Implementation**: As the model descends, detailed designs are created for individual modules (both mechanical and electronic) and software components.
- **Integration and Verification**: The right side of the Vee model represents the integration of subsystems and their verification against the original requirements.
- **Validation and Testing**: The final step involves validating the complete system in realistic conditions, such as field trials or simulated environments.

By following the Vee model, AUV developers can ensure that every design decision is traceable back to user requirements and that each module is verified and validated before final integration. This iterative process helps reduce errors and improve overall system reliability.

2.5.2 Requirements Mapping

A fundamental aspect of systems engineering is the mapping of user requirements to system functions. Tools such as the House of Quality (HoQ) and Quality Function Deployment (QFD) can be used to translate user requirements into engineering specifications. In AUV development, this process can be represented as follows:

- **Identifying Core Functions**: Determining the essential functions an AUV must perform—such as propulsion, sensing, navigation, communication, and power management.

- **Mapping to Subsystems**: Dividing these functions among modular hardware and software components. For example, the navigation function might be implemented by a combination of inertial measurement units (IMUs), GPS (when surfaced), and acoustic positioning systems (Aristizábal et al., 2021; Zuluaga et al., 2022).
- **Establishing Interfaces**: Defining standard interfaces and data protocols that enable communication between these subsystems. Standardisation ensures that modules developed by different teams can be integrated without extensive rework.

2.6 STANDARDS AND CERTIFICATION

The development and deployment of underwater vehicles, particularly those designed for specific industrial applications and environments, are often guided by rules and standards established by classification societies and various established engineering codes and standards.

2.6.1 Classification Societies

Organisations like ABS (American Bureau of Shipping) and DNV-GL (Det Norske Veritas—Germanischer Lloyd) provide comprehensive rules for the classification of underwater technology (UWT) systems. These rules cover aspects such as design, materials, manufacturing, construction, and testing.

ABS Rules for Building and Classing Underwater Vehicles, Systems, and Hyperbaric Facilities provide requirements for manned submersibles, unmanned underwater vehicles, and underwater working machines. Specific sections cover scope, conditions of classification, materials, pressure vessels, testing, and electrical systems, among others. These rules also define AUV and HAUV categorisation and approval paths.

DNV-GL Rules for classification for Underwater technology (UWT) summarise loads to be considered for design and specify principles for the construction of manned submersibles, including mating, control, and monitoring systems. These rules also cover pressure vessels and gas cylinders.

Classification society rules, such as those from ABS and DNV-GL, provide a foundational framework for ensuring the safety and structural integrity of underwater vehicles. These rules often mandate compliance with or consideration of other recognised standards for specific components and systems. The choice of a specific external standard would depend on component type, material, and intended operating conditions.

For instance, pressure vessels and viewports have specific, well-established standards like ASME PVHO-1 and PVHO-2, which are directly referenced in the classification rules. Materials are often specified according to national standards like ASTM or Classification Society-specific grades. Piping and electrical systems commonly refer to standards developed for marine vessels, adapted for the underwater

TABLE 2.3
Classification Rules and Standards for Underwater Vehicles

Standard/Rule	Scope/Application
ABS Rules for Building and Classing Underwater Vehicles, Systems and Hyperbaric Facilities	Requirements for design, materials, manufacturing, construction, and testing of manned submersibles, unmanned underwater vehicles, systems, and hyperbaric facilities. Defines UUV/AUV categories and approval paths.
DNV-GL Rules for classification: Underwater technology (UWT)	Summarises design loads for submersibles, UUVs, and underwater working machines. Principles for the construction of manned submersibles, including mating, control, and monitoring. Covers pressure vessels and gas cylinders.
ABS Rules for Conditions of Classification (Part 1A)	Requirements for conditions of classification for underwater vehicles, systems, and hyperbaric facilities. Specific sections referenced for documentation submission.
ABS Rules for Survey After Construction (Part 7)	Surveys after construction for handling systems. In-service inspection for hyperbaric chambers. Periodic surveys for classed AUVs/HAUVs.
ABS Rules for Alternative Arrangements, Novel Concepts and New Technologies (Part 1D)	Qualification and/or approval for new technologies utilised.
ABS Guidance Notes on Review and Approval of Novel Concepts	Provides details on the Approval-In-Principle (AIP) process for prototype/developmental vehicles during conceptual design.
ABS Guidance Notes on Risk Assessment Applications for the Marine and Offshore Industries	Methodology for detailed risk analysis, e.g., for battery installation (alternative: equivalent recognised national or international standards).
Society for Underwater Technology (SUT) Code of Practice for UUVs	Recommended code of practice (Not legally binding).

environment. Handling systems are subject to standards for lifting appliances and specific component standards for ropes and hooks.

A list of Classification Rules and Standards applicable to Underwater Vehicles is given in Table 2.3.

2.6.2 Legal and Certification Framework

Unmanned marine vehicles (UMVs or UXVs) have been in use for many years, but a formal and binding legal framework for safety has not yet been adopted. The growth in use of the technology has evolved under a loose legal framework, leading to self-regulation within the UMV community. A recommended Code of Practice for UUVs exists from the Society for Underwater Technology (SUT), but it is not legally binding. There is a need for an internationally agreed set of procedures to be developed and adopted for managing risks associated with AUV operations. Legal ambiguities exist, for instance, regarding whether an AUV is to be considered a 'ship' or 'scientific research installations or equipment' under UNCLOS, suggesting a need for specialised instruments or clarification.

Classification societies define processes for design review and certification. As an example, ABS rules outline various approval paths for AUVs/HAUVs based on size and capability, as follows:

- Micro, Small, and Medium AUVs/HAUVs may follow Product Design Assessment (PDA) or Type Approval (TA) paths.
- Large and Extra Large AUVs/HAUVs have options including I3P Verification (with Independent Review Certificate, IRC, and/or Certificate of Conformity, CoC) or full classification. Classification includes design review, new construction surveys, installation/integration surveys, and periodic surveys after construction.
- Prototype or developmental vehicles are handled case by case. Designers may apply for Approval in Principle (AIP) during the conceptual phase, demonstrating compliance with rules and standards.

2.6.3 Rules Applicable for Structures

Specific requirements apply to pressure hulls, especially for manned submersibles, in various Classification Society Rules.

- **Materials**: Materials for pressure hulls must be suitable for their intended purpose and processes like welding. Materials not covered by standard rules may be considered with agreement from the Society. Plate materials for pressure boundaries are to comply with specified ABS Hull Grades, US Navy Grades, or ASTM standards.
- **Analysis and Calculations**: Metallic pressure boundary components are to be analysed for external pressure, potentially using recognised national or international pressure vessel standards or codes not less effective than the rules. Design calculations for pressure hulls, including internal/external pressure and proofs of stress for load cases like Nominal Diving Pressure (NDP), Test Diving Pressure (TDP), and Collapse Diving Pressure (CDP), are to be submitted.
- **Tolerances**: Manufacturing tolerances for pressure hulls are prerequisites for applying strength calculations. Tolerances for weld sinkage and edge offset are specified. Dimensional tolerances and surface finish details are to be submitted.
- **Proof Testing**: Pressure hulls are subject to hydrostatic or external pressure tests. Strain gauge measurements are to be recorded, and a report

comparing calculated versus measured stresses is to be submitted to validate the design. Differences over 10% require technical justification.

- **Viewports**: Acrylic windows and viewport assemblies (window, flange, retaining rings, seals) are covered by specific rules. Design parameters for windows are subject to the latest edition of ASME PVHO-1. Operating temperature limits and pressurisation/depressurisation rates are specified. Design life for windows subject to tensile stresses is not to exceed ten years. Design certification, pressure testing certification, and fabrication certification for windows are required, often referencing ASME PVHO-1 forms.

Rules also apply to pressure vessels fitted to an underwater system other than the main pressure hull. Plans, calculations, and data for pressure vessels are to be submitted for review and approval by the respective Classification Society.

- **External and Internal Structures**: External structures (support frames, skids) and internal equipment/facilities are subject to rules. They are to be constructed following sound engineering practice and approved plans. Welded connections are to be adequately constructed and possible to access for inspection. Support frames/skids for PVHOs should be adequate for function and loads.

2.6.4 Standards for Design and Construction

Rules summarise loads to be considered for design. Environmental conditions are the basis for calculations. Designers and construction yards must prove actual loads in each special case, considering effective loads if they are higher than specified in the rules. Construction is to follow sound engineering practice and approved plans. Structural design certification requirements have been discussed. Examples of other aspects of design standards are given below, and these standards are summarised in Table 2.4.

- **Handling Systems**: Launch and recovery systems (LARS) for manned underwater units must have loose gear hooks equipped with a positively lockable latch. Winches must meet design requirements. Rope must be constructed according to a recognised standard. Motion compensation systems must be designed so that a single failure does not cause load loss, verified by risk analysis. Handling systems that are part of a classed system are to meet applicable requirements.
- **Piping Systems and Machinery**: Piping systems are to be designed, constructed, installed, inspected, and tested according to the (ABS) Marine Vessel Rules, except as modified by their UWT rules. Systems exposed to pressure higher than design pressure require safety relief devices. Manufacturer's specifications and capacities are to be submitted. Piping/tubing is to be mechanically protected.
- **Electrical Systems**: Equipment must be designed for the operating environment to minimise risks. Distribution panels should be accessible, and

TABLE 2.4
Standards/Codes for Specific Components and Systems

Component/System	Relevant Standard/Code	Application Details
Pressure vessels (general)	Recognised national or international pressure vessel standards or codes	Analysis for external pressure is not less effective than ABS rules
Pressure vessels (welded)	Applied design code or standard	Additional non-destructive tests (NDT) required
Materials (General)	ABS Rules for Materials and Welding (Part 2)	
Materials (plates for pressure boundaries)	ABS Hull Grades E, EH32, EH36; US Navy Grades HY-80, HY-100; ASTM A516 Grades 55, 60, 65, 70; ASTM A537 Class 1 & 2; ASTM A517 Grades A, B, E, F, J	Plate materials are to comply with one of these
Viewport assemblies & windows	ASME PVHO-1 (for design parameters)	Certification required for Design, Pressure testing and Fabrication
Piping systems	ABS Marine Vessel Rules	For design, construction, installation, inspection, testing
Piping systems	ABS Rules for Building and Classing Underwater Vehicles, Systems and Hyperbaric Facilities	As applicable
Electrical systems	Part 4, Chapter 8 of ABS Marine Vessel Rules	Electrical installations and equipment design, fabrication, installation, inspection, testing
Electrical systems	ABS Rules for Building and Classing Underwater Vehicles, Systems and Hyperbaric Facilities	General requirements for electrical equipment
Handling systems (launch and recovery)	ABS Section 20, 'Handling Systems'	
Handling Systems (launch and recovery)	DNVGL-ST-0377 (For manned units: DNVGL ST-0378 Lifting Appliances)	
Rope for handling systems	API Specification 9A, or US Federal Specification RR-W-410a	
Production equipment (hydrocarbon)	ABS Rules for Building and Classing Offshore Units/Facilities on Offshore Installations	Equipment for the production of hydrocarbons are to comply
Welding procedures and NDT	Various recognised standards	Welded pressure vessels require additional NDT beyond applied design code

power to chambers should be possible to disconnect. Cables/wiring within the boundary of the pressure hull are to be low smoke, low/zero halogen type cables. Emergency power sources are required.

2.6.5 Risk Assessment

A risk-informed approach sets prescriptive requirements for verification and validation activities. Risk assessment considers safety risks, failure consequences, and ability to regain control. Risks related to integration, function failure, equipment failure, and operational hazards are assessed.

Formal processes and regulatory regimes, although not mandatory for UUVs, are useful. ABS provides requirements for autonomous functions and for building and classing underwater vehicles. For example, a detailed risk analysis for battery installation is required per ABS Guidance Notes on 'Risk Assessment Applications for the Marine and Offshore Industries' or equivalent recognised standards.

Plans and data for risk assessment can be submitted for review. Procedural and legal risks exist, and therefore safety procedures and protocols for operation are crucial. Human error has been identified as a dominant source of faults. Emergency abort systems are included in vehicle mission logic to enable vehicle recovery in case of major systems failure.

2.6.6 Standards for Interoperability

While structural and mechanical design relies heavily on established engineering codes and classification rules, standards for communication and data transfer are still evolving, driven by the increasing complexity and connectivity of unmanned systems. The need for interoperability is pushing the adoption of common protocols and data formats across different platforms and agencies.

Standards like Joint Architecture for Unmanned Systems (JAUS) aim to provide a common language for unmanned systems regardless of domain, while others like ASTM F2595-07 are specific to UUV sensor data. The move towards service-oriented architectures and shared knowledge representations through ontologies reflects an effort to standardise the *interaction* and *information exchange* layer, enabling modular software and collaborative operations.

Various communication and data standards or protocols relevant for UUVs are listed in Table 2.5.

2.6.7 Choice and Application of Standards

The choice of standards is influenced by the vehicle's intended purpose (military, commercial, scientific), operating environment (depth, duration, autonomy level), and size. For example, military applications often require higher levels of robustness, security, and adherence to defence-specific standards like NATO STANAGs or CDL. Civilian commercial vehicles, while still requiring high reliability (especially in offshore energy), might prioritise cost-effectiveness and use widely recognised

TABLE 2.5
Communication and Data Standards/Protocols

Standard/ Protocol/Concept	Context/Application	Description/Purpose	Notes/References
Standard data formats for ISR products	Joint, network-centric environment	Employed to permit networked exploitation of data.	Needed to facilitate the exploitation of UAV data.
National Information Exchange Model (NIEM)	Multi-agency interoperability	Data exchange model.	Between government agencies.
NATO STANAG 4559	NATO, military ISR	Integrates ISR data.	
Joint Architecture for Unmanned Systems (JAUS)/SAE AS-4	Unmanned systems	Common message formats and data protocols. Migrated to an international industry standard (SAE AS-4).	Concepts used as a foundation for knowledge representation.
NATO STANAG 4586	NATO, unmanned aviation systems (UAS)	Unmanned aviation's counterpart to JAUS.	
Common Data Link (CDL) family	DoD unmanned systems	Waveform specifications for interoperability.	For standardising interfaces.
ASTM F2595-07	UUV sensor data	Standard guide for formats.	One of the initial UUV standards for data formats.
Generic Sensor Format (GSF)	Bathymetry	Data format.	For bathymetric data from sensors.
Bathymetric Attributed Grid (BAG)	Bathymetry	Data format standard.	For derived bathymetric information.
WMO, IOC, and similar standards	Data collection formats	For standardising UUV data collection formats.	Supported by a wide range of data processing software.
Compact Control Language (CCL)	REMUS vehicle, communication between AUVs	Used as a control language for communication between AUVs.	Originally designed specifically for REMUS with acoustic devices.
Service-Oriented Architectures (SOA)	System design, Interoperability	Modules provide services and advertise capabilities for information exchange.	Used in the ICA framework.

(Continued)

TABLE 2.5 (CONTINUED)
Communication and Data Standards/Protocols

Standard/ Protocol/Concept	Context/Application	Description/Purpose	Notes/References
Middleware (e.g., MOOS, OceanSHELL, CORBA)	Communication architecture	Network-based communication and message passing. CORBA used in the Open Control Platform (OCP) for distributed component communication.	Simplify application building and manage communication. Supports distributed heterogeneous components.
Ontologies (Foundation, Core, Application)	Knowledge representation, information sharing	Used in semantic frameworks for distributed knowledge representation and agent interaction.	Provides common machine understanding. Core ontology extends JAUS concepts.
MAC and Routing protocols	Underwater wireless sensor networks	Needed to manage data transmission and path selection in acoustic networks.	Design depends on channel characteristics.

industry standards. Scientific vehicles may incorporate cutting-edge sensors and require flexible data handling standards. The increasing autonomy of AUVs compared to ROVs necessitates more focus on standards for on-board decision-making, data processing, and autonomous communication protocols.

As mentioned earlier, UUVs/AUVs can be differentiated by size (Micro to Extra Large), and different approval paths may be specified within classification rules. While the rules become more stringent for larger vehicles, the specific application of *all* external standards (e.g., which particular communication protocol or material standard) is not explicitly mapped to each size category in the provided sources. Generally, the complexity of the vehicle and mission dictates the suite of standards applied. A small, simple AUV performing basic mapping might adhere to core structural and safety rules, basic navigation, and perhaps a standard data logging format, while a large intervention AUV or a military platform would require a much broader set of standards covering complex hydraulics, manipulation systems, advanced communication, interoperability protocols, and cyber resilience.

2.7 CONCLUSION

Design and development of AUVs start from identification of tasks, selection of equipment, developing alternative configurations, and refinement of the selected configuration. The refinement can take place in several stages, followed by testing and trials. Typical payload and layout of AUVs have been discussed. An outline has been provided of the major systems to be designed, which will be covered in detail in subsequent respective chapters.

Various approaches for initial sizing of AUVs have been outlined. Modular design provides flexibility to adapt AUVs to varied and challenging missions while ensuring that each component can be developed, verified, and upgraded independently. Applying systems engineering methodologies, such as the Vee model and iterative IV&V, ensures that design decisions are fully traceable to user requirements and that every subsystem is rigorously tested before final integration.

The various agencies and standards applicable for the design and construction of uncrewed subsea systems have been discussed. These include codes for structural design, inspection, equipment, components, and interoperability.

In the subsequent Chapters 3 to 8, we shall examine each major component and system of UUVs in further detail. We shall examine possible choices, typical equipment and approaches for their selection or development.

REFERENCES

Agha, G. (1990, September). Concurrent object-oriented programming. *Communications of the ACM, 33*, 125–141. https://doi.org/10.1145/83880.84528

Allmendinger, E. (1990). *Submersible Vehicle Systems Design.* Jersey City: Society of Naval Architects and Marine Engineers.

Ambastha, S., Das, S., Pal, D., Nandy, S., Shome, S., & Banerjee, S. (2014). Underwater Terrain Mapping with a 5-DOF AUV. *Indian Journal of Geo-Marine Sciences, 43*, 106–110. Retrieved May 2025, from https://www.researchgate.net/publication/271079407_Underwater_terrain_mapping_with_a_5-DOF_AUV

Aristizábal, L., Zuluaga, C., Rúa, S., & Vásquez, R. (2021, May 11). Modular Hardware Architecture for the Development of Underwater Vehicles Based on Systems Engineering. *Journal of Marine Science and Engineering, 9*, 516. https://doi.org/10.3390/jmse9050516

Autosar. (n.d.). *AUTomotive Open System ARchitecture.*

Busby, R. F. (1976). *Manned Submersibles.* Arlington: Office of the Oceanographer of the Navy.

Dieter, G., & Schmidt, L. (2021). *Engineering Design.* McGraw-Hill Education.

Fathallah, E., & Helal, M. (2016, April). *Optimum Structural Design of Deep Submarine Pressure Hull to Achieve Minimum Weight.* 11th International Conference on Civil and Architecture Engineering. Cairo.

Geyer, R. A. (1977). *Submersibles and their Use in Oceanography and Ocean Engineering.* New York: Elsevier.

Hwang, J., Bose, N., Millar, G., Gillard, A., Nguyen, H., & Williams, G. (2014). Enhancement of AUV Autonomy Using Backseat Driver Control Architecture. *International Journal of Mechanical Engineering and Robotics Research, 10*, 292–300. https://doi.org/10.18178/ijmerr.10.6.292-300

INCOSE. (2015). *INCOSE Systems Engineering Handbook: A Guide for System Life Cycle Processes and Activities.* Wiley.

Jackson, H. A. (1992). Fundamentals of Submarine Concept Design. *SNAME Transactions, 100*, 419–448.

Madhan, R., Desa, E., Prabhudesai, S., Sebastiao, L., Pascoal, A., Mascarenhas, A., . . . Khalap, S. (2007). *Mechanical Design and Development Aspects of a Small AUV - Maya.* IFAC Conference on Manoeuvring and Control of Marine Craft. Retrieved May 2025, from http://drs.nio.org/drs/handle/2264/702

Moore, S. W., Bohm, H., & Jensen, V. (2010). *Underwater Robotics: Science, Design and Fabrication.* Monterey, CA: Marine Advanced Technology Education Center.

Moos-Ivp. (n.d.). *Retrieved from moos-ivp.* org.

NASA. (2017). *NASA Systems Engineering Handbook.*

Ros. (n.d.). *Retrieved from ros.* org.

Ullman, D. (2017). *The Mechanical Design Process.* David Ullman LLC.

Zuluaga, C., Aristizábal, L., Rúa, S., Franco, D., Osorio, D., & Vásquez, R. (2022, March 25). Development of a Modular Software Architecture for Underwater Vehicles Using Systems Engineering. *Journal of Marine Science and Engineering, 10*, 464. https://doi.org/10.3390/jmse10040464.

3 Form, Propulsion, and Powering

3.1 HYDRODYNAMIC DRAG AND HULL FORM

The body of the AUV is referred to as 'hull' (external and internal), and its external shape and size are referred to here as 'form'. The optimal hull form is an engineered balance between hydrodynamic efficiency, hydrostatic stability, propulsion compatibility, and mission practicality.

The hull form and its structure are designed based on the following considerations:

- Provide sufficient internal volume for components and payloads.
- Minimise hydrodynamic drag.
- Withstand hydrostatic pressure at design depth through adequate structural integrity.
- Ensure stability both statically (maintaining orientation) and dynamically (during manoeuvres).
- Integrate with propulsion systems.
- Offer modularity for sensors and power systems.

In realistic design scenarios, competing objectives such as range (maximisation), drag (minimisation), and payload volume (maximisation) must be reconciled through multi-objective optimisation. These dependencies are represented in Figure 3.1.

3.1.1 Hydrodynamic Drag

The external hull form is primarily based upon the diameter and length of the pressure-proof enclosures that it contains. A smaller vessel will have lower drag, but internal volume would be driven by considerations of stored energy and payload. For greater maximum speed, it is more efficient to increase power plant capacity rather than trying to reduce displacement.

The total drag (or 'resistance') experienced by an AUV is a combination of frictional drag, form drag, wave-making drag (when on the surface), and induced drag due to appendages (such as fins, antenna projections, etc).

Skin friction drag arises from viscous shear along the wetted surface. This component of resistance is reduced with minimising surface area. For a given internal volume, a spherical shape has the minimum surface area, and hence the minimum skin friction drag.

DOI: 10.1201/9781003737476-3

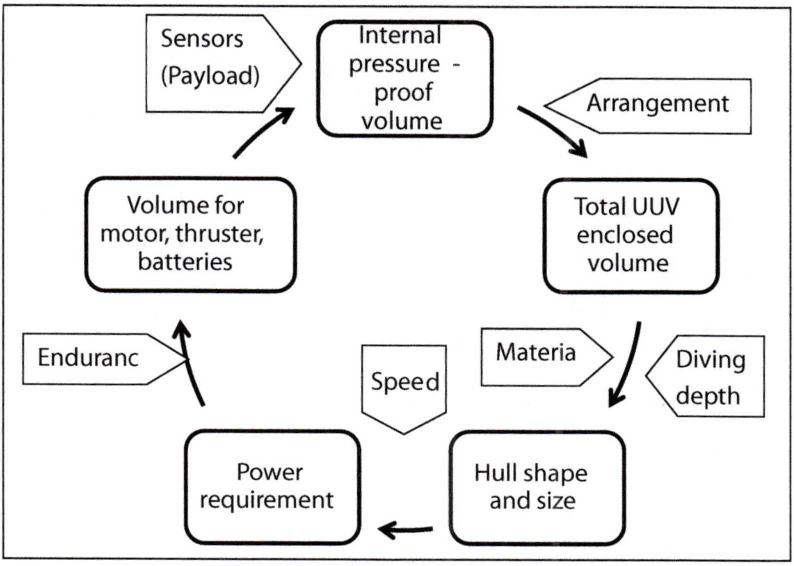

FIGURE 3.1 Typical considerations (dependencies) and iterations for designing the hull form of an AUV.

Source: **Author.**

Pressure resistance (or 'form drag') results from flow separation and pressure differentials across the hull. A needle-like form will have the minimum pressure drag. Adopting a shape with a teardrop profile reduces flow separation and thus minimises pressure drag.

By modifying the parameters characterising the vessel shape, it is useful to examine the impact of varying the form on the drag and power requirements (Dell & Rand, 2001). AUVs typically operate at low speeds, i.e., low values of Reynolds number (although beyond laminar flow); hence skin friction drag is expected to be more significant than form drag. Therefore, more attention should be paid to minimising the total exposed surface area (i.e., wetted surface area) of the vessel and any projections and appendages, rather than trying to optimise its overall form.

3.1.2 Axisymmetric Form

The advantage of using an essentially axisymmetric form is that it is possible to geometrically estimate the form volume, and thus estimate the resistance and power requirements using empirical methods. The methodology given by Jackson (1992) for axisymmetric submarine-like forms is easy to implement and adequately accurate for initial estimates. An example of axisymmetric form used for typical AUVs is shown in Figure 3.2.

FIGURE 3.2 Jaya AUV by CSIR-NIO Goa.

Source: **Author.**

One of the major parameters for the form drag is the hull's length-to-diameter (L/D) ratio, which affects both hydrodynamic drag and manoeuvrability. Ratios between 6:1 and 8:1 favour laminar flow and low drag for higher-speed vehicles. Shorter ratios (3:1 to 5:1) enhance turning ability and stability for slower, survey-type AUVs. The choice of L/D ratio therefore needs to balance slenderness (for form drag reduction) with interior volume needs, manufacturing constraints, and minimising surface area (for skin friction drag reduction) (Jackson, 1992).

AUVs usually operate at low speeds (3 to 5 knots, i.e., about 1.5 to 2.5 m/s) (*Bluefin Robotics Unmanned Underwater Vehicles,* n.d.). Low speed is preferred for better performance of sensors and for minimising power consumption, since power requirements increase steeply with speed (approximately proportional to the cube of speed). At such low speeds, it is the frictional drag component that dominates, for which it is more beneficial to minimise the external surface area of the AUV rather than streamlining its shape. A cylindrical hull with fairings at both ends is the typical form selected for AUVs, and its drag coefficient can initially be estimated from empirical data accordingly, which can be refined by CFD analysis (Hong, Wang, & Zhang, 2024).

Therefore, typically, AUV hulls are axisymmetric bodies of revolution with a teardrop or torpedo-like geometry. The pressure hull consists of one or more cylindrical or spherical structures designed to resist external pressure efficiently, while minimising weight.

3.1.3 Non-axisymmetric Forms

A square cross-section affords greater utilisation of internal space, but is less efficient in withstanding external pressure. Driven by payload requirements, large AUVs are seen being designed with square cross sections to maximise internal volume. This presumably allows modular or containerised payloads such as mines to be carried or

FIGURE 3.3 'Manta Ray' large UUV demonstrated by Northrop-Grumman.

Notes: **https://www.northropgrumman.com/what-we-do/mission-solutions/sensors/manta-ray**

deployed during the mission. This would be efficient in terms of weight if the box-like form is mainly free-flooding (i.e., not pressure-resistant), enclosing one or more pressure-resistant spheres housing electronics, and the rest of the volume exposed to seawater.

3.1.4 Biomimetic Forms

Biomimetic design draws inspiration from marine animals that have evolved to move efficiently underwater.

Computational Fluid Dynamics (CFD) analyses of biomimetic forms show that whale or shark-inspired geometries produce lower viscous resistance and improved flow attachment compared to traditional torpedo hulls. Hybrid designs merge laminar flow optimisation with flexible fin-based propulsion for environment-specific missions.

Rigid hull forms inspired by biological shapes can be used as a 'lifting wing'. An example is the 'Manta Ray' UUV developed by Northrop Grumman (Figure 3.3), which is derived from a DARPA project of the 2000s for buoyancy-driven underwater gliders.

3.2 HYDROSTATICS AND HYDRODYNAMICS

3.2.1 Weight and Buoyancy Considerations

The weight and buoyancy need to be accurately estimated and matched for a UUV (or submarine) to be able to dive and surface whenever intended. Imbalances are likely to occur due to inaccurate or incomplete data of vehicle components during design, variations during assembly or installation, and variations in water density.

The major components of the weight of a typical AUV are its batteries and its pressure-resistant structure. The greatest contributors to volume are the enclosed volume of the pressure hull and the volume of water displaced by any large components outside the pressure hull (e.g., external pressure-compensated batteries).

Imbalance between weight and buoyancy of a vessel can be rectified by adding weights or by providing (rigid) buoyancy foam. The former option is always preferable due to the relative ease of adding/removing weights, provided this is catered for early in the design. Use of buoyancy foam material (which needs to retain its characteristics even at the rated depth of the vehicle) is a more difficult option to implement, since the shapes of their pieces must be customised and are not easily modified. The material compressibility needs to be evaluated carefully, particularly for very deep-diving AUVs.

It is also essential that the centre of gravity and centre of buoyancy of the UUV, when neutrally buoyant in water, are in the same vertical line, thereby maintaining a horizontal (level) attitude. This is shown in Figure 3.4. Adjustable weights, ballast systems or syntactic foams are used to achieve fine-tuned neutral buoyancy.

The disposition of weights should ensure that the transverse position of the centre of gravity (CG) of the vehicle is lower than its centre of buoyancy (CB). This is represented in Figure 3.5. Ensuring the CG is below the CB (by at least about 5 to 10 cm) results in a stable restoring moment, so that the vehicle remains upright when submerged.

It is common practice to design AUVs to be slightly positively buoyant by default, so that in case of loss of power, the vehicle would float up to the water surface. This

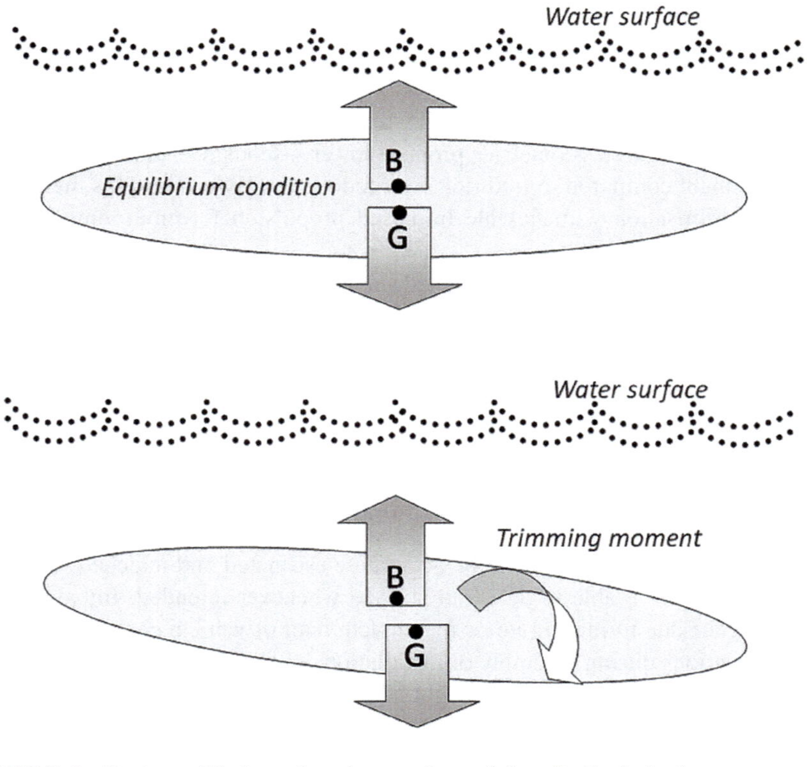

FIGURE 3.4 Static equilibrium of a submerged vessel (longitudinal view).

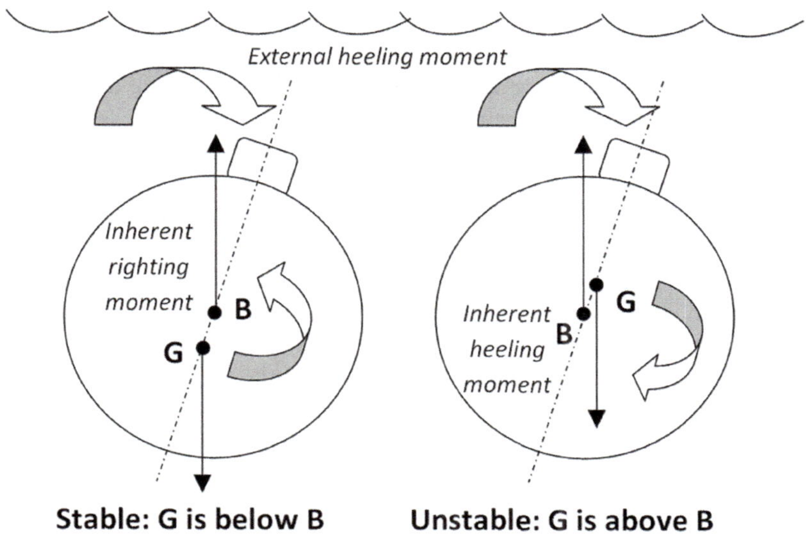

FIGURE 3.5 Static stability in heeling for a submerged vessel (cross-sectional view).

approach requires the vehicle to consume power to dive, either using vertical thrusters or (more commonly) by application of control forces through its hydroplanes, which are effective only when the vehicle has forward speed.

3.2.2 Manoeuvrability

For an autonomous underwater vehicle (AUV) or an unmanned underwater vehicle (UUV), achieving effective manoeuvrability is central to its operational performance. The hydrodynamic configuration of the bare hull must be refined not only for low resistance but also for controllability across six degrees of freedom. This is accomplished by integrating control surfaces (fins or planes) located at the stern, bow, or occasionally along the mid-body of the vehicle.

Control surfaces are aerodynamic appendages, such as rudders for yaw control (via differential pressure on port/starboard fins) and hydroplanes (elevators/stabilisers) for pitch and heave control. These control surfaces, such as rudders, elevators, or stern planes, are designed to deflect under actuator command, producing hydrodynamic moments that alter the vehicle's orientation. When the lateral control fins are deflected, they generate yawing moments enabling turning in the horizontal plane, while vertical fins produce pitching moments for ascent or descent. The geometry of these appendages—span, sweep, aspect ratio, and airfoil profile—governs both their lift generation and associated drag. Often, their initial sizing is based on empirical scaling laws drawn from torpedo hydrodynamics and further refined through computational fluid dynamics (CFD) simulations or tow-tank experiments.

For control surfaces to be effective in influencing the vehicle's motion, the vehicle must have sufficient speed. To generate sufficient dynamic lift, the forward velocity should exceed stall speed, typically $V > 0.3\ L^{0.5}$ m/s, where L is vehicle length in metres.

Linear stepper motors serve as high-precision actuators for fin deflection, with angles up to ±30°. Suitable placement of control surfaces within the propulsor slipstream augments low-speed controllability by increasing local flow velocity, thereby enhancing lift coefficients C_L by 20–50%.

The motion of a UUV is inherently nonlinear and coupled. Its mathematical model represents the interplay between hull geometry, mass distribution, control surface characteristics, and added-mass hydrodynamic effects. Thus, the mathematical model of motion of the vehicle correlates its form (shape) with the location and shape of its control surfaces, speed, weight (and CG), and the control forces (deflection of fins) with the resultant orientation and trajectory of the AUV in the three-dimensional underwater domain.

In a subsequent chapter (Chapter 7), we shall see how the guidance, navigation, and control are linked to this mathematical model of motion of the vehicle.

3.2.3 Power Estimation

The hydrodynamic resistance is linked to propulsion power as follows:

$$\text{Power required} = \text{Drag} \times (\text{Speed/Propulsive efficiency})$$

Optimising hull geometry minimises required shaft horsepower (SHP).

Propellers are fixed- or variable-pitch rotors. In simplest terms, when described via actuator disk theory, thrust $T = \rho\, A\, v_e\, (v_e - v_0)$ relates motor power $P = \eta\, \rho\, A\, v_e^{\,3}/2$ to diameter D, advance ratio $J = v_0/(nD)$, and efficiency $\eta_p \approx 0.7\text{–}0.85$.

For propeller design, the pitch-diameter ratio (P/D) optimises torque Q and cavitation number $\sigma = (Ps - Pb)/(0.5\rho n^2 D^2)$. Nozzle ducts (Kort-type) boost thrust by 30–50% at low J (< 0.5).

3.2.4 Iterative Hull Form Optimisation

An integrated hydrostatic-hydrodynamic design approach ensures that the hull maintains structural and buoyant equilibrium at rest while minimising dynamic drag during motion. Iterative design cycles would involve the following typical steps:

- Determining pressure hull thickness and material.
- Computing buoyancy balance through volume distribution.
- Estimating drag by empirical methods.
- Selection of propulsion components.
- Estimate of manoeuvrability (directional stability).
- Re-evaluating centre of gravity shifts due to internal payload rearrangement.
- Performing CFD-based flow and resistance tests for finalised configuration.

Several iterative cycles would typically be necessary to refine the hull to achieve neutral buoyancy, static stability, and optimal flow dynamics.

3.3 AUV PROPULSION SOLUTIONS

Propulsion solutions for AUVs can be classified as:

- Conventional thrusters.
- Buoyancy-driven gliding.
- Biomimetic undulatory/jet mechanisms.
- Wave-actuated kinematics.

Propulsion modalities exhibit differentiated performance envelopes in endurance, specific energy consumption (Wh/km), acoustic signatures, and agility metrics (Bradley et al., 2001).

3.3.1 Propulsion Approaches

- **Thrusters**: These are propeller-like devices (usually integrated with a propelling motor) that can generate force along a fixed direction relative to the vehicle body. Multiple fixed thrusters are often used to control the required DOFs for a mission. Vectored thrusters or nozzles can change the direction of the generated thrust, enhancing manoeuvrability.
- **Buoyancy-Driven Propulsion**: Gravity and buoyancy changes within a vehicle can be used for slow, energy-efficient movement. Underwater gliders use variable buoyancy systems to achieve a saw-tooth dive profile. Hydrodynamic wings convert the vertical motion into forward glide. This method is exceptionally energy-efficient, enabling missions that span weeks or even months (Graver, 2005).
- **Biomimetic Propulsion**: Systems that imitate biological swimmers, such as fish tails or fins. Aims to improve efficiency, low-speed manoeuvrability, and reduce noise compared to conventional propellers (Bandyopadhyay, 2005). Examples include undulating fins/bodies or flapping tails.

3.3.2 Thruster-Based Propulsion

Thrusters produce propulsive force through momentum transfer to the surrounding fluid. Basic concepts of propeller design apply, with the shroud or nozzle surrounding the blades providing additional protection from damage (Griffiths, 2003).

Prevalent configurations encompass paddle-wheel thrusters for distributed thrust generation, electrically actuated thrusters employing brushless DC (BLDC) motors, tunnel thrusters optimised for lateral (sway) and vertical (heave) manoeuvres at velocities below 0.5 m/s, and water-jet propulsors that expel high-velocity fluid streams for cavitation-resistant operation. These elements fundamentally govern vehicle dynamics, including maximum speed, acceleration profiles, and manoeuvrability margins.

AUV thrusters predominantly use brushless DC (BLDC) motors directly coupled to a propeller. Electric motors supply rotational torque to thrusters and actuators, with BLDC variants preferred for their high power density (up to 5 kW/kg), low cogging torque, and efficiency exceeding 90% across 20–80% load. Brushed DC motors offer simplicity but suffer brush wear; hydraulic motors excel in tethered applications, accepting pressurised fluid at 100–300 bar. Overload protection integrates current sensors triggering at 150% rated torque, preventing thermal runaway.

Thrusters usually feature flooded stators for hydrostatic pressure equalisation (up to 100 m depth) and seawater lubrication, obviating the need for dynamic seals. Examples include Blue Robotics T200 (5 kgf max at 16V, 17–32 A) and T500 (16.1 kgf at 24V, >1 kW) (*Blue Robotics Thrusters, Actuators, and Lights*, n.d.).

Advantages include rapid response time (<0.1 s), enabling quick acceleration and fine manoeuvring. Thrusters are modular and scalable, and can be arranged around the vehicle to provide control in all six degrees of freedom, if necessary for the mission. Scalable thruster arrays achieve $T_{net} > 2\,\rho\, V_d L^2\, C_d$. They are a mature technology, and commercially available with extensive field data.

The major limitation of thrusters is that they are energy intensive, and their continuous operation involves high power consumption. Typical power density ~500–1000 W/kg yields $E_{\text{spec}} > 100$. Compared to other propulsion options, thrusters are noisier, which may affect sensitive sensors. Cavitation inception (typically at $\sigma < 0.3$) may generate broadband noise 60–80 dB re 1 μPa @1m, but cavitation may not be of concern at higher ambient pressure (greater depth).

More than one thruster mounted on an underwater vehicle can allow directional control without the use of lift-dependent fins (control surfaces). This may be advantageous for low-speed vehicles, or for missions where hovering (position-keeping) is essential.

A typical thruster model is shown in Figure 3.6. Selection and sizing of thrusters are discussed in greater detail in a subsequent section.

FIGURE 3.6 Blue Robotics Thruster T500.

Notes: **Blue Robotics: https://bluerobotics.com/product-category/thrusters/**

3.3.3 Buoyancy-Driven Gliders

These vehicles essentially keep diving and surfacing as a consequence of controlled changes in their buoyancy and pitch angle, thereby resulting in their net forward motion, referred to as a 'saw-tooth' profile of trajectory. Well-known buoyancy gliders developed in the 2000s in US universities include the Slocum, Seaglider, and Spray (Ray, Singh, & Seshadri, 2011). An example is shown in Figure 3.7.

Their low energy consumption enables missions of range more than 1,000 km, consuming less than 10 Wh/km. Minimal power usage allows for missions lasting weeks to months. The absence of continuously running motors also results in quiet operation (noise levels <40 dB re 1 μPa @1m). However, average speeds are quite low. For example, the Slocum achieves an average speed of 0.68 knots, which increases to 2 kn with auxiliary thrusters. Values for the Seaglider are similar (*Seaglider Autonomous Underwater Vehicle*, n.d.).

Buoyancy engines regulate net buoyancy for dive/surface cycles or depth holding. Change in vehicle buoyancy is achieved by employing variable ballast tanks (VBT) with piston pumps displacing 0.1–1% vehicle volume, or electrochemical engines generating gas. Trim systems of the vehicle change (or maintain) pitch angle θ by translating internal masses (e.g., batteries at 10–20% displacement) or via coupled control surfaces, stabilising via moment arm shifts.

The change in buoyancy is a function of the density of surrounding seawater and the variable volume ($\Delta B = \rho\ g\ V_b\ \Delta\phi$). Variable buoyancy engines (piston or oil-bladder) fitted in the vehicle change their density ($\Delta\rho/\rho \approx 0.1\%$) and centre of gravity. In combination with a change in pitch angle, via shifting weights or by high-aspect

FIGURE 3.7 'Seaglider' Underwater Glider of APL, University of Washington.

Notes: **Ray, Singh, and Seshadri (2011); https://apl.uw.edu/project/project.php?id=seaglider**

wings (Aspect Ratio > 5), this results in saw-tooth trajectory profiles of the vehicle. Typical glide ratios (horizontal distance to vertical distance covered) vary from 10:1 to 20:1.

A limitation is that typical speeds of buoyancy-driven gliders are much lower than comparable thruster-driven AUVs. Their net cruising speed (V_{cruise}) <0.3 m/s, which restricts their usage to transect surveys. They also have limited agility, with a path curvature radius of more than 100 m, which precludes agile station-keeping.

3.3.4 Wave-Generated Motion

Some unmanned surface vehicles (USVs) exploit ocean wave energy. The Wave Glider, for example, utilises the relative motion between a surface float and a submerged glider to produce thrust without fuel consumption (*WaveGlider: Persistent Ocean Monitoring*, n.d.). This method offers extended endurance and sustainable operation. An example is shown in Figure 3.8.

FIGURE 3.8 'Wave Glider' developed by Liquid Robotics.

Note: **liquid-robotics.com**

Dual-body USVs (surface float + submerged foil) harness wave orbital velocities via phase-locked heaving/pitching, producing thrust $T_w = \rho\, Aw\, \zeta_a\, \omega^2$ without prime movers. Wave Glider supplements with solar auxiliary power (~50 W).

Their advantage lies in near-infinite endurance in waves with significant wave height greater than 0.5 m. However, their performance is highly environment-dependent. Their variation in thrust with sea state can be more than 50%. Also, they are slower and their manoeuvrability is limited, with a rate of turn typically less than 5 degrees per minute.

3.3.5 Biomimetic Propulsion

Biomimetic systems replicate the efficient swimming mechanisms of marine animals—such as the caudal fin motions of fish or jet propulsion in squids—to potentially offer superior efficiency and manoeuvrability with lower noise. Although many biomimetic designs remain in the experimental phase, they are promising for applications that demand high agility and stealth (Li et al., 2023).

Common references include dolphins, whales, rays, and sharks. These natural swimmers achieve minimal drag and high agility through streamlined bodies, compliant surfaces, and flexible appendages (Roberts & Sutton, 2013). Examples of such inspiration include the following:

- **Cetaceans (Whales, Dolphins)**: Exhibit fusiform (torpedo-like) shapes optimising drag reduction.
- **Elasmobranchs (Sharks, Rays)**: Utilise boundary layer control through microstructures (riblets) that minimise skin friction.
- **Fish and Eels**: Employ undulating bodies and caudal fins for efficient thrust and manoeuvring.

Realising biomimetic hulls requires flexible yet pressure-resistant materials (Duchi, Viveros, & Burciaga, 2023). Emerging technologies, such as soft robotics and 3D-printed composites, could potentially allow smooth shape morphing without compromising integrity (Hossain et al., 2025). Advantages include very low noise levels (<50 dB) and thrust augmentation.

Mimicking biological locomotion can yield quieter and more efficient systems, with potential for superior manoeuvrability in complex flow conditions, as well as in multi-domain operations such as beaches, swamps, or shallow water with a muddy bottom. They can develop adaptive gait via reinforcement learning. However, their actuation involves high complexity (10+ DOF linkages), and they are still at the development stage, TRL 4–6. There are scaling challenges beyond 1 m in length.

3.4 THRUSTER-BASED PROPULSION DESIGN

Since thrusters are the most widely used propulsion method, this section looks at details of their design, performance, and control (Wadoo & Kachroo, 2011; Hasan, 2017).

3.4.1 Design Principles

Conventional thrusters employ sensor-less BLDC motors with fully flooded housings that allow water circulation for cooling and lubrication (Moore, Bohm, & Jensen, 2010). This design not only enhances pressure tolerance but also reduces maintenance requirements by eliminating traditional seals. While many systems use a direct mechanical coupling between the motor and propeller, recent developments include magnetic or reconfigurable couplings that permit thrust vectoring without rotating the entire assembly.

Sensor-less BLDC motors (slot/pole optimised for minimal cogging, e.g., 9-slot/12-pole) in flooded housings can sustain 10 bar/cm immersion via balanced internals. Direct couplings yield motor efficiency of more than 95%. Magnetic gears enable ±15° vectoring at 1 Nm torque.

3.4.2 Performance Parameters

Key performance metrics for thrusters include:

- **Power Range**: Small units may operate in the 50–500 W range, while larger thrusters can require several kilowatts.
- **Voltage and Current**: Typical operating voltages range from 7 V to 24 V (with some high-power units operating at higher voltages). For example, the T200 thruster draws approximately 17–32 A, whereas the T500 draws around 43.5 A at 24 V.
- **Thrust Output**: Thrust scales as $T \propto \rho n^2 D^4$. Expressed in kgf or newtons, thrust levels vary from a few kgf (for small AUVs) to over 16 kgf in high-power models.
- **Coupling and Noise**: Direct coupling is standard; however, advanced magnetic couplings can offer dynamic thrust vectoring. Fully flooded designs help minimise cavitation and mechanical noise.

Table 3.1 summarises the specifications of typical commercially available thrusters for AUV/ROV applications (*Thrusters: Product information*, n.d.).[1]

3.4.3 Control Systems

Electronic speed controllers (ESCs) use pulse-width modulation (PWM) to manage power delivery, ensuring efficient motor operation across a range of loads. Modern systems often integrate ESCs with on-board sensors to adjust parameters in real time for optimal performance and to safeguard against faults. Field-oriented control (FOC) ESCs can regulate currents via PWM (10–20 kHz), and sensor fusion (IMU/encoder) yields speed hold within ±1%.

TABLE 3.1
Typical Commercial Thruster Specifications

Thruster Model	Operating Voltage	Power Range	Current Rating	Thrust Output	Coupling Mechanism	Key Features
Blue Robotics T200	7–20 V	~205–645 W	~17 A (12 V)–32 A (20 V)	~3.7–6.7 kgf (8.2–14.8 lbf)	Direct, fully flooded	Compact, affordable BLDC design; widely used in platforms like BlueROV2
Blue Robotics T500	16–24 V	~1 kW+	~43.5 A at 24 V	~16.1 kgf (35.5 lbf)	Direct, fully flooded; patented design	High-power with ~3 × T200 thrust, rugged; pressure tolerant
Tecnadyne AUV Thrusters	Typically 12–48 V	50 W–11 kW	Varies (10s to 100s of A)	From 10s to 100s of kgf	Direct coupling; often with redundant seals	Engineered for work-class AUVs; robust in harsh conditions

3.4.4 Thruster Selection

Choosing the correct thruster involves careful consideration of vehicle characteristics, mission requirements, and environmental conditions. An integrated approach combining empirical estimation, CFD analysis, bench testing, and performance modelling is recommended to optimise thruster selection. When selecting thrusters, apart from meeting the peak speed (thrust) requirement, the following aspects should also be considered:

- **Match to Vehicle Dynamics**: The thruster's thrust curve should align with the vehicle's acceleration and deceleration needs. For instance, an AUV requiring rapid course corrections may benefit from thrusters with a low response time and high efficiency at partial loads.
- **Efficiency Over Operating Range**: Evaluate efficiency curves to ensure that the thruster operates near its peak efficiency during most of the mission.
- **Physical Integration**: Size and weight constraints are critical. Thrusters must fit within the vehicle's design envelope without compromising buoyancy or balance.
- **Environmental Considerations**: Choose thrusters with suitable pressure ratings and corrosion-resistant materials, especially for deep-water or harsh marine environments.
- **Control and Redundancy**: For mission-critical applications, consider systems with built-in redundancy and advanced control capabilities (e.g., magnetic-coupling units that offer thrust vectoring).

A systematic selection process involves the typical steps shown in the flowchart in Figure 3.9. This approach ensures that the propulsion system is neither over-designed (resulting in unnecessary energy consumption and weight) nor under-powered for the vehicle's mission profile (Rutherford, 2008).

Broad guidelines for selecting a thruster from a product catalogue can be further simplified and summarised as follows:

- Consider drag (hydrodynamic) loads [$F_d = 0.5\ \rho\ V^2\ C_d A$] for maximum speed as well as typical operating speed (with some margins).
- Estimate the required thrust, [$T_{req} = F_d(1+\mu)$] with μ = 1.5 (as an example) for redundancy.
- Propulsive efficiency [$\eta_p(J)$] of the selected thruster(s) should be greater than 70% of the operating regime.
- Increase in vehicle drag due to thruster integration should be less than 5% after mounting on the vehicle.
- Pressure rating of material (Ti/Al alloy) of the thruster should be about 1.5 times the depth rating of the vehicle.

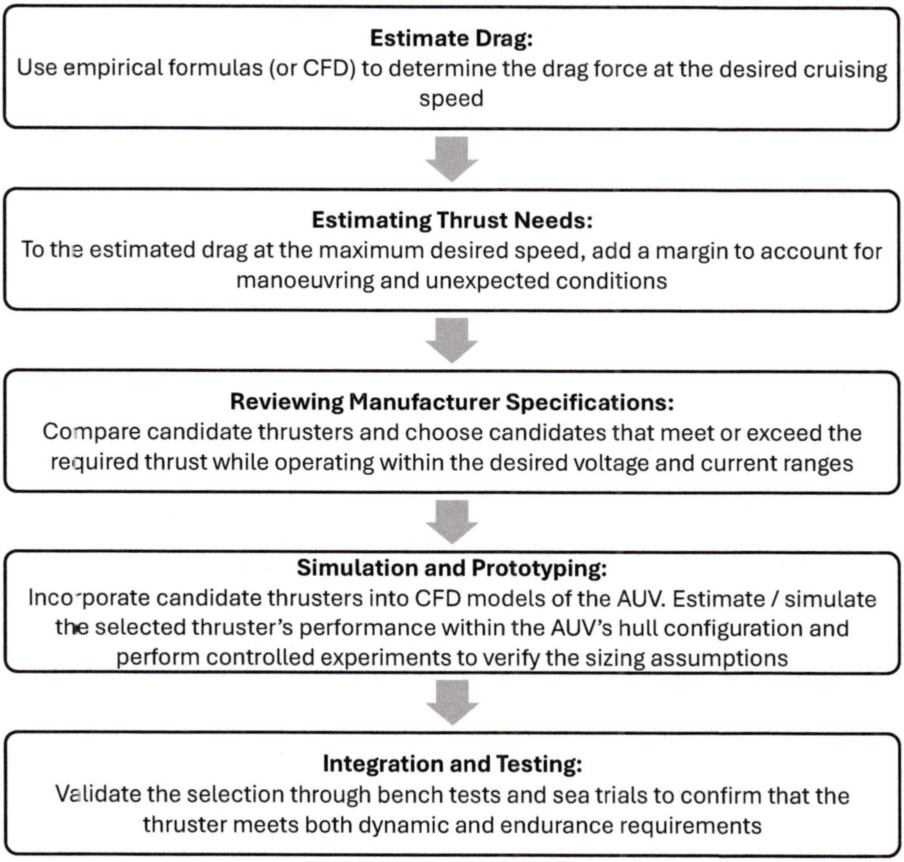

FIGURE 3.9 Flowchart for thruster selection.

Source: **Author.**

3.5 CONCLUSION

AUV hull form design intertwines hydrodynamic theory, hydrostatic stability, propulsion integration, and increasingly, biological inspiration. Effective designs employ axisymmetric streamlined geometries (L/D 6–8), minimise drag via laminar flow maintenance, and incorporate adaptive propulsion alignment.

Conventional thrusters, with their proven responsiveness and modularity, continue to be the preferred solution for applications requiring agility and speed. In the future, advances in CFD, AI-driven optimisation, and novel materials will further refine thruster-based systems and accelerate the integration of hybrid propulsion solutions. Such innovations are expected to enhance energy efficiency and expand the operational capabilities of AUVs.

Other solutions, such as buoyancy-driven gliders, wave-powered systems, and emerging biomimetic approaches, offer advantages in energy efficiency and stealth. Biomimetic hulls show promise for reducing viscous resistance, enhancing manoeuvrability, and mitigating noise. Future developments will likely merge flexible biomimetic surfaces with real-time morphing capabilities, enabled by smart composite materials.

NOTE

1. https://tecnadyne.com/thrusters

REFERENCES

Bandyopadhyay, P. (2005, January). Trends in Biorobotic Autonomous Undersea Vehicles. *IEEE Journal of Oceanic Engineering*, *30*, 109–139. https://doi.org/10.1109/joe.2005.843748

Blue Robotics Thrusters, Actuators, and Lights. (n.d.). Blue Robotics. Retrieved May 2025, from https://bluerobotics.com/product-category/thrusters/

Bluefin Robotics Unmanned Underwater Vehicles. (n.d.). General Dynamics Mission Systems. Retrieved May 2025, from https://gdmissionsystems.com/underwater-vehicles/bluefin-robotics

Bradley, A. M., Feezor, M. D., Singh, H., & Sorrell, F. Y. (2001). Power systems for autonomous underwater vehicles. *IEEE Journal of Oceanic Engineering*, *26*(4), 526–538.

Dell, R. M., & Rand, D. A. (2001). *Understanding Batteries*. Royal Society of Chemistry. https://doi.org/10.1039/9781847551065

Duchi, E. A., Viveros, B. I., & Burciaga, S. P. (2023). Design of Modular Underwater Vehicle Biomimetic Hull Towards Onboard Lab-On-a-Chip Devices. *International Journal of Mechanical Engineering and Robotics Research*, *12*, 324–331. https://doi.org/10.18178/ijmerr.12.5.324-331

Graver, J. (2005). *Underwater Glider: Dynamics, Control and Design*. Princeton University Press.

Griffiths, G. (Ed.). (2003). *Technology and Applications of Autonomous Underwater Vehicles*. Taylor & Francis. https://doi.org/10.1201/9780203023249

Hasan, H. (2017). *Underwater Robotics: Science, Design and Fabrication*. Malaysia: UTM Press: Malaysia.

Hong, L., Wang, X., & Zhang, D.-S. (2024, August). CFD-based Hydrodynamic Performance Investigation of Autonomous Underwater Vehicles: A Survey. *Ocean Engineering*, *305*. https://doi.org/10.1016/j.oceaneng.2024.117911

Hossain, G. A., Saimoon, N. Z., Salehim, M. N., Ahmed, M. M., & Masud, M. H. (2025, July). Recent Advancements in Biomimetic Application in Underwater Glider Development: Current Opportunities and Future Trends. *Ain Shams Engineering Journal*, *16*(7). https://doi.org/10.1016/j.asej.2025.103410

Jackson, H. A. (1992). Fundamentals of Submarine Concept Design. *SNAME Transactions*, *100*, 419–448.

Li, G., Liu, G., Leng, D., Fang, X., Li, G., & Wang, W. (2023). Underwater Undulating Propulsion Biomimetic Robots: A Review. *Biomimetics*, *8*(3), 318. https://doi.org/10.3390/biomimetics8030318

Moore, S., Bohm, H., & Jensen, V. (2010). *Underwater Robotics: Science, Design and Fabrication*. Monterey, CA: Marine Advanced Technology Education Center.

Ray, A., Singh, S., & Seshadri, V. (2011). Underwater Gliders: Force Multipliers for Naval Applications. *Warship 2011: Naval Submarines and UUVs*. Bath, UK: Royal Institution of Naval Architects. https://doi.org/10.3940/rina.ws.2011.03

Roberts, G. N., & Sutton, R. (Eds.). (2013). *Further Advances in Unmanned Marine Vehicles*. The Institution of Engineering and Technology. https://doi.org/10.1049/PBCE002E

Rutherford, K. T. (2008). *Autonomous Underwater Vehicle Design Considering Energy Source Selection and Hydrodynamics*. University of Southampton.

Seaglider Autonomous Underwater Vehicle. (n.d.). Applied Physics Laboratory (APL), University of Washington. Retrieved May 2025, from https://apl.uw.edu/project/project.php?id=seaglider

Thrusters: Product information. (n.d.). Tecnadyne. Retrieved May 2025, from https://tecnadyne.com/thrusters

Wadoo, S., & Kachroo, P. (2011). *Autonomous Underwater Vehicles: Modeling, Control, Design, and Simulation*. CRC Press Taylor & Francis Group.

WaveGlider: Persistent Ocean Monitoring. (n.d.). Liquid Robotics. Retrieved May 2025, from https://www.liquid-robotics.com/wave-glider/how-it-works/

4 Energy Storage and Management

In harsh subsea environments, the stored energy of UUVs is usually the limiting factor for the length of their missions such as ocean exploration, environmental monitoring, offshore infrastructure inspection, and defence applications conducted without human intervention. The energy available on board constrains mission duration, range, and the quality and quantity of data that can be collected (Moore, Bohm, & Jensen, 2010). Therefore, the design, selection, and management of energy sources and power systems for AUVs are of great significance.

4.1 CHALLENGES IN SUBSEA POWERING

The challenges in the areas of AUV energy and propulsion that make it a challenging engineering problem are summarised below.

- **Limited On-board Energy**: Due to space, weight, and buoyancy constraints, only a limited amount of energy can be carried on board. This restricts mission duration and range.
- **High Propulsion Demands**: The energy required for propulsion increases with the cube of the speed. Therefore, any high-speed manoeuvres can rapidly deplete batteries. Improving the overall energy efficiency of the vehicle system, from the power source conversion to the propulsor's interaction with the water, is essential.
- **Harsh Environmental Conditions**: The underwater environment is corrosive, cold (or warm), and under high pressure at depth. Energy storage devices and power electronics must be designed or selected to operate reliably under these conditions.
- **Weight and Volume Constraints**: Sensors, batteries, and power distribution systems all add mass and occupy volume. Maintaining neutral buoyancy and ensuring hydrodynamic efficiency requires careful trade-offs.
- **Integration Complexity**: Power distribution must be managed across multiple subsystems with varying voltage and current requirements. In addition, many AUVs must include energy management systems that monitor the state of charge, temperature, and overall health of the energy source in real time. Self-monitoring, diagnosis, and remediation are essential for autonomous systems.
- **Safety**: Adhering to safety standards for batteries and other high-energy systems, especially concerning fire hazards or handling requirements.

 DOI: 10.1201/9781003737476-4

- **Cost**: High-performance energy sources and propulsion systems can be costly. Operational costs are also significant, including launch, recovery, and maintenance.

4.2 ENERGY CONSUMPTION COMPONENTS

AUV energy consumption can be broadly categorised into three main components: propulsion, payload, and hotel load.

4.2.1 Propulsion Power

Propulsion is typically the most energy-intensive part of an AUV's operation. Propulsion power is the energy (P_P) required to move the vehicle through the water. It is often the largest energy consumer, especially on long missions. Propulsion power increases with the cube of velocity ($P_P \propto U^3$) for a given drag coefficient and surface area. Factors like vehicle shape, drag, and environmental conditions (currents, waves) significantly affect propulsion power (Griffiths, 2003).

The power required is influenced by:

- **Hydrodynamic Drag**: Generally increases with the cube of the velocity.
- **Efficiency of the Propulsion System**: This includes the motor, gear train (if any), and propeller. Propeller efficiency, transmission losses, and motor control all affect overall power consumption.
- **Operating Speed**: A balance must be struck between speed and energy consumption. Most AUV missions are designed to operate at low speeds (e.g., 1–3 knots) to maximise endurance.

Typical propulsion loads for mid-range AUVs vary widely. High-speed manoeuvres may require power levels in the tens of kilowatts (e.g., 20–50 kW) during short bursts (Hasan, 2017). The majority of operations are at a lower cruising speed, requiring power in the range of 1–5 kW, depending on vehicle size and hydrodynamic design (Griffiths, 2003; Roberts & Sutton, 2013).

4.2.2 Hotel Loads

Hotel loads include the continuous energy demand of electronic systems that maintain control, communication, and sensor operation. The overall hotel load in many AUV designs is typically in the range of 50–100 W (when averaged over the mission), although peak loads during sensor operation or data transmission can be higher (AUV ROV Sensors, n.d.; Roberts & Sutton, 2013).

Hotel loads can also be considered as the power required to operate all non-propulsion vehicle systems (P_V). This includes vehicle equipment operation and onboard processing requirements.

- **Vehicle Equipment**: Control planes, buoyancy systems, navigation equipment, modems, communication systems, fault monitoring sensors, vehicle computer, payload computers.
- **On-board Processing**: Running control architectures, algorithms (planning, decision-making, fault tolerance), and processing sensor data. This can demand a high duty cycle from the payload computer, impacting endurance.

4.2.3 Mission/Payload Loads

This category is the power required for vehicle mission sensors (P_{MS}) and payload components (P_{PC}). These denote power drawn by mission-specific equipment such as imaging systems (still/video cameras), chemical sensors, sonars (side-scan, bathymetric, multibeam, sub-bottom profilers), magnetometers, and environmental sensors (e.g., CTD, diffused oxygen).

Power consumption varies greatly by sensor type; for instance, Synthetic Aperture Sonar (SAS) is power-hungry. Visual observation (cameras) is also power-intensive. Variables affecting the load due to sensors would include their duty cycles, resolution/sampling settings, and logic for operation/storage/communication.

While accounting for these loads, double counting with hotel loads should be avoided.

4.2.4 Typical Energy Loads

The total energy (E) required for a mission can be estimated by the sum of the power consumption of these components over the mission duration, which is related to range (R) and velocity (V): $E = (P_P + P_V + P_{MS} + P_{PC}) \times (R \,/\, V) \,/\, 3600$ kWh.

Table 4.1 summarises typical energy loads of various AUV systems.

4.3 POWER GENERATION AND DISTRIBUTION

Power is distributed across the AUV via a centralised DC bus, with DC-DC converters used to supply the correct voltage to each subsystem (Bradley et al., 2001). Key considerations for power distribution architecture include the following:

- **Isolation and Redundancy**: To protect sensitive systems from voltage spikes or faults in other parts of the system.
- **Efficiency**: High-efficiency converters are essential to minimise losses, especially given the limited energy resources on board.
- **Scalability**: Modular power distribution systems allow new subsystems to be added with minimal redesign.

TABLE 4.1
Typical Energy Loads on AUVs

Subsystem	Power Requirement (W)	Duty Cycle	Remarks
Propulsion (cruising mode)	1–5 kW	Continuous during transit	Dominant power consumer; varies with speed and drag
Propulsion (high-speed burst)	20–50 kW (for short durations)	Brief bursts during manoeuvres	Peak power demands during acceleration or evasion
On-board computing and control	5–30 W	Continuous (can be throttled down)	Includes navigation, control algorithms, and monitoring
Environmental sensors (CTD, DO, pH)	0.1–1 W per sensor	Intermittent duty (e.g., 10–30% active)	Often duty-cycled to conserve energy
Acoustic sensors (active sonar)	5–20 W	Intermittent during active scanning	Power varies with ping rate and intensity
Optical imaging (HD camera)	1–5 W (low-light mode)	Continuous or triggered based on need	Higher for high-resolution video; may be duty-cycled
Communication (acoustic modem)	1–5 W	Variable (higher during data transmission)	Often includes standby and active modes
Additional payload (e.g., chemical sensors)	0.5–2 W each	Typically low duty cycle	Specific to mission requirements

4.3.1 Power Generation/Storage

The most common power source for AUVs is its on-board batteries. In recent designs, the focus has shifted towards increasing energy density and reducing the weight and volume of battery packs. Some of the options are mentioned below.

- **Battery Pack Configuration**: Battery packs are arranged in series or parallel (or both) to achieve the desired voltage and capacity (Dell and Rand, 2001). Modern systems use high-discharge Li-ion or LiPo cells to meet peak power demands.
- **Fuel Cells**: Fuel cells offer an alternative with a high energy density and the advantage of fast 'refuelling' (or fuel replacement) compared to recharging batteries. These systems are being considered for long-duration missions.
- **Hybrid Systems**: Hybrid solutions use a combination of batteries and supercapacitors. The batteries supply continuous energy, while the supercapacitors handle high-power bursts (such as for propulsion surges), thereby protecting the battery from stress.

4.3.2 Power Distribution

Once the energy is generated or stored, it must be distributed to the various subsystems in the AUV:

- **DC Power Bus**: Most AUVs use a DC bus architecture where the battery output is regulated and distributed through DC-DC converters to the different modules (propulsion, sensors, computing, etc.).
- **Power Management Electronics**: These systems monitor battery state of charge, temperature, and health. They also control load prioritisation and implement low-power modes during standby (Bradley et al., 2001).
- **Redundancy**: In safety-critical applications (especially in military AUVs), power distribution systems are designed with redundancy. Multiple power channels and backup batteries ensure that a failure in one channel does not jeopardise the entire mission.

Table 4.2 summarises typical AUV power generation and distribution systems.

4.4 ENERGY STORAGE CHARACTERISTICS

Different energy storage technologies offer varying performance characteristics critical for AUV design.

- **Energy Density vs. Specific Energy**: Both aspects are important. Energy density (Wh/L) affects vehicle volume requirements, while specific energy (Wh/kg) affects vehicle weight.
- **Power Density**: This is the power output per unit volume (W/L or W/dm^3) or mass (W/kg). The parameter is important for missions requiring high instantaneous power, like sprinting or high-power sensors. High power and high energy tend to be mutually exclusive for batteries.
- **Cycle Life**: This denotes the number of charge-discharge cycles a rechargeable battery can withstand. This is important for long-term operational cost and reliability.
- **Safety**: Batteries and fuel cells present safety considerations, such as fire hazards (Lithium) or handling requirements (semi-fuel cells).
- **Operating Temperature**: Affects performance and may require temperature control systems.
- **Cost**: Varies significantly among technologies and may need to be accounted for in terms of life-cycle expenses for a holistic comparison.

4.5 ENERGY STORAGE SOLUTIONS

Energy storage is a fundamental limiting factor for AUV endurance and mission capability. A variety of technologies are employed, each with its own characteristics, advantages, and disadvantages.

TABLE 4.2
Overview of AUV Power Storage, Generation, and Distribution

System Type	Description	Typical Voltage Range	Distribution Methods	Advantages	Disadvantages
Direct battery power	Supply DC power directly to subsystems	12–48 V typical	DC bus with DC-DC converters	Simplicity; mature technology	Limited by battery capacity; weight constraints
Fuel cell-based systems	Generate electricity continuously, often in hybrid with batteries	24–48 V (as configured)	Integrated power electronics, DC bus with conversion	Extended endurance; fast refuelling possibilities	Complex integration; higher cost
Hybrid systems (battery + super-capacitor)	Combines high-energy batteries with supercapacitors for peak load management	12–48 V typical	Power management system with energy buffering	Smoothens transient load; extends battery life	Increased system complexity
Wireless power transfer	Underwater charging via magnetic induction or acoustic coupling	Variable	Dedicated charging coils/docking interface	Potential for in situ recharging; reduces downtime	Currently low efficiency; short range; still in R&D

4.5.1 Electrochemical Batteries

Batteries are the primary energy source for most UUVs. They store potential chemical energy and convert it to electrical energy. Battery performance is often evaluated by specific energy (Wh/kg) and energy density (Wh/L) (Dell and Rand, 2001).

4.5.1.1 Primary Batteries

These are designed for a single discharge. Examples are as follows:

- **Manganese Alkaline Cells**: They have been used in some AUVs (e.g., Autosub), offering a practical, low-cost option for less intensive use. Energy density is lower than that of lithium types. Performance is affected by current drain and temperature.
- **Lithium Primary Batteries:** They offer particularly high energy density. Lithium sulphur dioxide (~300 Wh/kg) and lithium thionyl chloride (~661 Wh/kg) are mentioned. They have been used in smaller vehicles like the REMUS AUV for a significant range.

4.5.1.2 Secondary (Rechargeable) Batteries

These are intended to be recharged after discharge. Examples are given below.

- **Lead-Acid Batteries**: These are well-established and widely used on submarines, but rarely used in modern AUVs, except where cost is an overriding factor and weight is less critical. They are relatively low-cost, but also have relatively low energy density (~25 Wh/kg). Sealed lead-acid (VRLA) cells have been used in some AUVs and sensors, offering reliability and being non-ferromagnetic, but they are heavy. Lead-Acid batteries are commonly used for engineering trials and short missions. Volume and weight increase significantly for extended cruising. They are robust and tolerant of harsh conditions, although they require maintenance and have limited cycle life.
- **Silver-Zinc Batteries**: They have much higher energy density (~120 Wh/kg) than lead-acid, but are expensive and have a shorter cycle life (40–100 cycles). They have been traditionally used for torpedoes and underwater vehicles.
- **Nickel-Metal Hydride (Ni-MH) Batteries**: They have higher energy density than nickel-cadmium (although lesser than lithium-ion batteries), but self-discharge can be an issue. NiMH batteries can be seen in older or budget-constrained designs; they are less common in modern high-end AUVs. They are robust and tolerant to overcharge conditions.
- **Lithium-Ion (Li-Ion) Batteries**: They are widely used due to high energy density (90–150 Wh/kg, 150–200 Wh/L). They have low self-discharge rates and relatively low weight. They have been used successfully in AUVs like Autonomous Benthic Explorer (ABE), Urashima, and REMUS. Most modern AUVs (e.g., REMUS 300, Bluefin-21) use Li-ion batteries,

sometimes in conjunction with pressure-tolerant packaging to minimise weight (*Bluefin Robotics Unmanned Underwater Vehicles*, n.d.). They offer longer cycle life (600–1,000 cycles) and can be pressure balanced. They have been associated with safety concerns (fire hazard), although these concerns are less critical for unmanned platforms. They require sophisticated battery management systems (BMS) and are sensitive to temperature and overcharge/over-discharge.

- **Lithium-Polymer Batteries**: They offer high energy density (130–190 Wh/kg, 170–240 Wh/L) and long cycle life (300–3000 cycles). Solid state design allows operation at ambient pressure, not requiring a pressure hull. They have been used in vehicles like Autosub6000 and PICASSO-1. LiPo batteries are often used in small or micro-AUVs, and for applications where design flexibility and weight savings are critical. They have similar energy density to Li-ion but can be manufactured in flexible, lightweight, and thin formats. They are more sensitive to mechanical damage and puncture. They require careful handling and charging protocols, and have a lesser cycle life compared to Lithium-Ion chemistry.
- **Metal Seawater Batteries**: These are based on magnesium or aluminium anodes. They can produce high energy densities (100–160 Wh/kg) but at low power densities.
- **Liquid Sodium Batteries**: Zebra batteries use a liquid sodium metal anode, operate at high temperatures (270–300°C), achieving energy densities around 100 Wh/kg for complete systems. They require insulation and preheating.

4.5.2 Common AUV Battery Types

Batteries are the most common energy source for AUVs because of their relative simplicity, high energy density, and maturity of the technology. The major battery chemistries used in AUVs are summarised in Table 4.3, (Figure 4.1).

4.5.3 Fuel Cells and Semi-fuel Cells

Fuel cells convert chemical energy directly into electrical energy through a chemical reaction. They generally offer higher energy density potential for long endurance missions compared to batteries.

4.5.3.1 Fuel Cells

In 'true' Fuel Cells, reactants are stored externally and fed to the cell. These can be of two main types:

- **Proton Exchange Membrane Fuel Cells (PEMFC)**: These are suitable for underwater applications, operating at lower temperatures (~70°C). The reaction product is pure water, which must be stored. They require storage of hydrogen and oxygen, which adds weight and volume to the system, and

TABLE 4.3
Commonly used AUV Battery Types

Battery Chemistry	Key Features	Example AUVs/Projects Using This Chemistry
Lithium-ion (Li-ion)	High energy density, mature technology	HUGIN AUV (Kongsberg), Bluefin-12, Odyssey IIx, REMUS series
Lithium-polymer (LiPo)	Flexible form factor, moderate energy	Small drones, lightweight underwater vehicles, various tech demos
Lithium titanate oxide (LTO)	Ultra-long cycle life, fast charging	Emerging in specialised deep-sea or military AUVs, research prototypes
Lithium sulphur (Li-S)	Very high theoretical energy density	Under research, experimental long-endurance AUV concepts
Alkaline aluminium/ hydrogen peroxide semi-fuel cell	Seawater activated, long endurance	CLIPPER AUV project by Woods Hole Oceanographic Institution
Seawater-activated batteries	Single-use, high-energy density	Disposable expendable underwater vehicles, sensors
Solid-state batteries	Safety improved, stable over temp/pressure	Experimental and under development for rugged marine applications
Hybrid battery-supercapacitor	High power bursts with energy storage	Military and high-power sonar AUV prototypes
PEM fuel cells with artificial gills	Use oxygen from seawater	Experimental autonomous systems for long-duration AUV missions

adds to complexity for refuelling. Closed-cycle systems are needed underwater with no emissions. They provide high energy density and longer operational durations and can be refuelled more rapidly than batteries can be recharged. Polymer electrolyte membrane (PEM) fuel cells have been used in the *Urashima* AUV (Japan).

- **Direct Methanol Fuel Cells**: These consume a dilute solution of methanol and water, producing CO_2 and water. They operate at low temperatures, but efficiency can be low, and by-product handling is a challenge.

4.5.3.2 Semi-fuel Cells

In these, one reactant is stored externally and fed to the reaction cell. In this category, Aluminium-Hydrogen Peroxide cells have demonstrated operational reliability, used in the HUGIN II and HUGIN 3000 AUVs. They offer high energy density, but require careful handling and replacement of reactants/anode. Practically available energy density is around 400 Wh/kg.

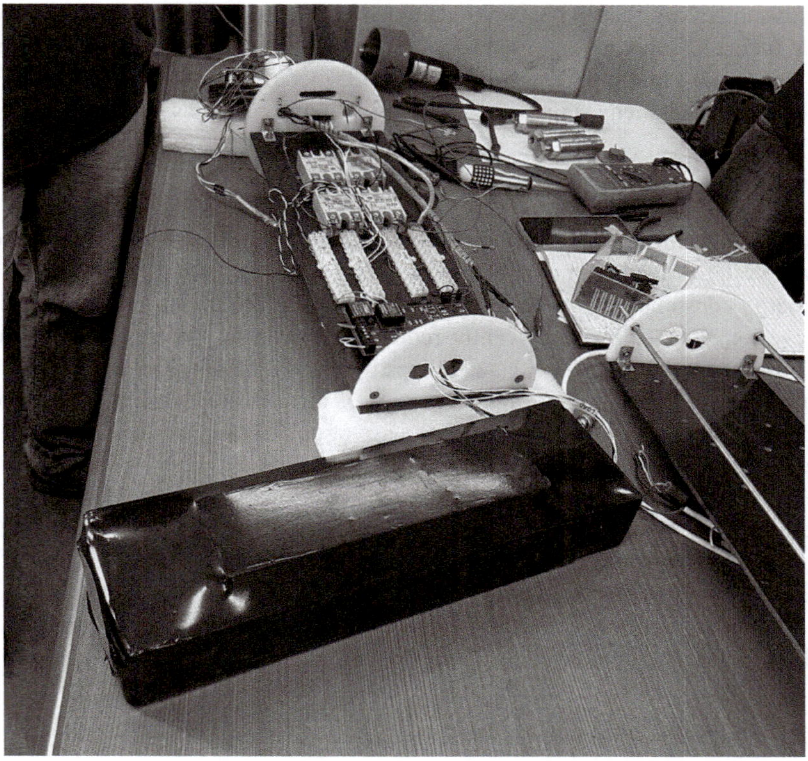

FIGURE 4.1 Lithium-ion battery pack being integrated in a small AUV.

Source: **NIO Goa.**

4.5.4 Hybrid Systems

Hybrid systems are at the research and development stage and are intended to combine batteries with fuel cells or supercapacitors. They are being tried out on advanced AUV designs targeting both high peak performance and extended endurance (Massaglia, 2021). According to DARPA, advanced UUV power supply technology based on Hybrid Fuel Cells would be capable of operating up to four hours at maximum power and operating under ideal conditions for 1,680 hours (Owens, 2022).

In a hybrid fuel cell system, a hydrogen fuel cell is supported by a traditional battery. The hydrogen fuel cell would use polymer electrolyte membrane (PEM) technology, where a combination of cathodes and anodes separates the protons and electrons of the hydrogen molecules and the PEM forces the electrons to travel through a circuit to generate power. The smaller, traditional battery would offer supplemental power when required by the size of the payload. Meanwhile, the hydrogen fuel cell would provide power at a stable rate for the steady load and recharge the small battery during low power intervals (Navy SBIR, 2015).

Advantages are that the batteries can provide high peak power for manoeuvres, while fuel cells supply long-term energy. Supercapacitors can handle transient loads and regenerative braking. Disadvantages include increased system complexity. There are integration challenges in managing multiple energy sources simultaneously.

4.6 ENERGY MANAGEMENT STRATEGIES

Energy management involves balancing power demands while optimising the energy storage system so that mission objectives can be met.

Energy consumption is a key constraint for vehicle endurance and range. Gliders achieve long ranges by significantly reducing propulsion power (lowest P_P term) and minimising hotel load (P_V), sometimes below 1W. Minimising hotel load often requires sacrifices in the type and frequency of measurements taken. Energy management algorithms can help optimise the use of on-board power sources and minimise energy use during task execution.

Heat generation is another aspect of energy consumption, particularly from batteries during discharge and from high-density electronics. Efficient cooling or thermal management is needed, especially in confined spaces.

Effective energy management is vital for maximising the operational efficiency and endurance of an AUV. Strategies include both hardware-based approaches and intelligent software control.

4.6.1 Duty Cycling and Load Scheduling

Many AUV subsystems do not require continuous operation. For example:

- **Sensor Duty Cycling**: Environmental sensors can be switched on intermittently, based on predefined sampling intervals or triggered by environmental changes.
- **Variable Propulsion Modes**: An AUV might operate in low-speed, energy-saving 'survey' mode and then switch to high-speed bursts only when necessary.
- **Standby Modes**: Non-critical systems can enter a standby state to conserve energy during long transit periods.

4.6.2 On-board Energy Monitoring

Modern AUVs employ sophisticated battery management systems (BMS) that continuously monitor voltage, current, temperature, and state of charge. They provide real-time data to the mission control software, allowing adaptive adjustments to power allocation. They can also alert operators to potential issues and automatically adjust system performance to conserve energy.

Battery management systems (BMS) provide real-time feedback to the AUV's control system, enabling:

- **State-of-Charge Estimation**: Accurate estimation of remaining battery capacity.
- **Thermal Management**: Active regulation of battery temperature through insulation, passive cooling, or even active liquid cooling.
- **Fault Detection**: Early identification of battery or power distribution faults, allowing the AUV to take corrective action.

4.6.3 Power Optimisation Algorithms

AUVs are beginning to use adaptive algorithms that adjust the power distribution based on real-time measurements and mission phases. Energy management algorithms play a critical role in reducing overall power consumption. They should dynamically allocate power to subsystems based on mission phase and load demands (Rutherford, 2008). Such adaptive power management should include the following features:

- **Predictive Energy Management**: On-board software can predict energy consumption based on mission profiles (historical data) and real-time sensor inputs and adjust operating parameters (speed, sensor activation, etc.) to optimise battery life.
- **Optimise Mission Profiles**: Modify speed, path, and operational modes dynamically to minimise energy use while still achieving mission objectives.
- **Adaptive Control**: Incorporate machine learning to improve efficiency over time, learning from previous missions to better schedule sensor activation and propulsion bursts.
- **Dynamic Load Adjustment**: Non-critical systems can be throttled or shut down during periods of high energy demand.

4.6.4 Thermal Management

Battery performance and longevity are highly dependent on temperature. Thermal management strategies include the following:

- **Insulation**: Battery compartments are often insulated to maintain optimal operating temperatures.
- **Active Cooling**: In some designs, small pumps or fans circulate coolant around the batteries.
- **Thermal Coupling**: Using waste heat from propulsion systems or on-board electronics to warm batteries in cold environments.

4.7 ENERGY HARVESTING

While batteries remain the predominant energy source, other technologies are emerging and are expected to mature in the years to come. Greater breadth of mission profiles for current and future naval UUVs requires longer endurance stealthy propulsion systems that extend the current capability of 10–40 hours to several days or weeks. Current and future anticipated technologies based solely on high-energy-density batteries may not provide adequate endurance for such missions.

This section reviews the major modalities of underwater energy harvesting, describing the technologies that are currently at the forefront of research and development. Energy harvesting technologies aim to capture ambient energy from the underwater environment, supplementing on-board energy reserves.

4.7.1 Wave Energy Conversion (WEC)

Wave energy converters are devices designed to capture the kinetic and potential energy of ocean waves. In underwater applications, devices are often moored to the seabed or integrated into floating structures. The energy extraction mechanism can involve oscillating water columns, point absorbers, or attenuators that deform in response to wave motion. Devices (often based on piezoelectric or electromagnetic induction principles) convert the kinetic energy of water motion into electrical energy (*WaveGlider: Persistent Ocean Monitoring*, n.d.). They can be useful for AUVs operating in areas with strong currents or wave action. Such systems might not supply high power but can extend mission duration by trickle-charging on-board batteries (*WaveGlider: Persistent Ocean Monitoring*, n.d.).

4.7.1.1 Underwater WEC Devices

Recent advances in underwater WEC systems have focused on miniaturisation and robustness. For example, compact point absorber designs have been tested to power AUVs and sensor platforms. These systems typically consist of a buoyant float coupled to a hydraulic or electromagnetic power take-off (PTO) system that converts mechanical motion into electrical energy. Laboratory experiments have demonstrated energy conversion efficiencies of up to 40% under ideal wave conditions, with prototypes being tested in coastal regions (Wang, 2022).

4.7.1.2 Challenges in Wave Energy Harvesting

Harvesting systems must be robust, lightweight, and efficient at converting low-frequency, low-amplitude energy fluctuations. Key challenges include the following:

- **Irregular Wave Patterns**: The stochastic nature of ocean waves requires robust control algorithms to optimise energy extraction.
- **Structural Fatigue**: Continuous mechanical stress can lead to wear and tear on moving components.
- **Biofouling**: Marine organisms adhering to device surfaces can reduce efficiency over time.

4.7.2 Tidal and Current Energy Harvesting

Tidal and current energy converters harness the kinetic energy of moving water. Unlike wave energy, tidal currents are highly predictable, making them attractive for long-term power generation. Turbines designed for underwater use capture the energy of water flowing past their blades (*Ocean Energy Storage*, n.d.).

Several turbine designs have been developed specifically for underwater environments:

- **Horizontal-Axis Turbines**: Similar to wind turbines, these are mounted on seabed structures. Recent research has focused on optimising blade profiles for low Reynolds number flows.
- **Vertical-Axis Turbines**: These devices can capture energy from water flowing in any direction and are often easier to maintain.
- **Biomimetic Designs**: Inspired by aquatic organisms, some designs mimic the shape and movement of fish fins to improve efficiency in turbulent conditions.

Field demonstrations have shown that underwater turbines can achieve power densities of up to 2–3 kW/m^2 under optimal tidal conditions. Innovations in materials and coatings have significantly enhanced the durability of turbine blades, reducing maintenance requirements and extending operational life (Teeneti et al., 2021).

4.7.3 Thermal Energy Harvesting

4.7.3.1 Ocean Thermal Energy Conversion (OTEC)

OTEC systems exploit the temperature difference between warm surface water and cold deep water. Conventional OTEC plants are large-scale and designed to provide grid-level power. However, scaled-down versions for underwater applications are under active research.

4.7.3.2 Thermoelectric Generators (TEGs)

TEGs convert heat directly into electricity using the Seebeck effect. In underwater settings, TEGs can be deployed on the exterior of submersibles or on fixed structures near thermal vents. Recent developments include the use of nanostructured materials to improve the thermoelectric figure of merit (ZT), thereby enhancing efficiency. Laboratory tests have achieved conversion efficiencies in the range of 5–8% for prototype devices, with ongoing research aimed at increasing this figure (Yang et al., 2026).

The primary challenges in underwater thermal harvesting are the following:

- **Low Temperature Gradients**: In many parts of the ocean, the temperature differential is modest, limiting the available energy.
- **Material Constraints**: Finding materials that are both efficient and resistant to long-term underwater corrosion is critical.
- **System Integration**: Efficient heat exchangers are needed to maximise energy capture without disrupting the local marine environment.

4.7.4 Salinity Gradient Energy Harvesting

Salinity gradient energy, often referred to as 'blue energy', exploits the chemical potential difference between freshwater and saltwater. When these two types of water mix, energy is released that can be captured using technologies such as reverse electro-dialysis (RED) or pressure retarded osmosis (PRO).

Recent experiments have focused on optimising membrane technology and reducing energy losses.

- **Reverse Electro-dialysis**: This method uses ion-selective membranes to create an electrical potential between saltwater and freshwater compartments. Current research is addressing issues of membrane fouling and efficiency.
- **Pressure Retarded Osmosis**: PRO systems use a semi-permeable membrane to generate pressure differences that drive turbine generators. Advances in membrane material science have improved system efficiency and longevity.[1] However, long-term operation in corrosive environments can lead to fouling and degradation.

4.7.5 Vibration and Pressure Fluctuation Harvesting

4.7.5.1 Piezoelectric Energy Harvesters

Piezoelectric materials generate an electrical charge in response to mechanical stress. In underwater applications, they can be integrated into structures that experience continuous vibrations, such as oceanographic sensor arrays or seabed installations. Advances in piezoelectric composite materials have led to devices that are more efficient and robust against mechanical fatigue.

4.7.5.2 Triboelectric Nanogenerators (TENGs)

TENGs convert mechanical energy into electrical energy via contact electrification and electrostatic induction. They are particularly suited to harvesting energy from irregular, low-frequency motions typical of underwater environments. Recent designs incorporate flexible materials that can conform to curved surfaces and are resistant to biofouling. Experimental setups have demonstrated power outputs sufficient for low-power sensor networks, with the added benefits of scalability and low cost (Teeneti et al., 2021).

Although promising, piezoelectric and triboelectric systems generally produce low power densities and are best suited for powering small-scale devices (Yang, 2026). Durability is another challenge, since continuous exposure to mechanical stresses and corrosive seawater requires advanced materials that can maintain performance over long operational periods.

4.7.6 Solar Energy Harvesting

Solar Powered AUVs (SAUVs) require surfacing daily to recharge batteries via photovoltaic (PV) arrays. They offer potentially limitless energy for long endurance but are limited by available sunlight, biofouling of panels, and the need to surface. Increasing the array size can increase drag.

Solar panels integrated on AUVs that surface periodically can harvest solar energy. Although sunlight attenuates rapidly underwater, solar cells can be effective during surface intervals (Teeneti et al., 2021). Gliders and other AUVs that spend a significant amount of time near the surface can benefit from solar recharging. The solar panels must be robust, waterproof, and optimised for variable light conditions.

4.7.7 Implementation of Energy Harvesting

Integrating underwater energy harvesting technologies into practical systems involves addressing multiple engineering and environmental challenges. This section outlines key considerations for system design, integration, and deployment.

4.7.7.1 Materials and Durability

The underwater environment is harsh. Saltwater corrosion, biofouling, and pressure extremes necessitate the use of specialised materials:

- **Corrosion-Resistant Alloys**: Stainless steel, titanium, and advanced composites are often used for structural components.
- **Anti-fouling Coatings**: Surface treatments that resist biological growth help maintain efficiency over time.
- **Flexible and Robust Polymers**: For transducer elements in piezoelectric or triboelectric systems, polymers must withstand repeated deformation without degradation.

4.7.7.2 Power Management and Storage

Harvested energy must be efficiently captured, stored, and managed. Integration with energy storage systems such as advanced batteries or supercapacitors is critical (Liu et al., 2024). Power Conditioning requires rectifiers, voltage regulators, and maximum power point tracking (MPPT) algorithms to ensure that harvested energy is converted and stored optimally. Maturity of underwater power transfer technologies could potentially extend the range and endurance of subsea underwater systems (Mohsan et al., 2022). Hybrid Systems that combine multiple energy harvesting modalities can provide redundancy and continuous power even when one source is intermittent. An example is shown in Figure 4.2.

4.7.7.3 Deployment Strategies

The success of underwater energy harvesting depends not only on the efficiency of individual devices but also on their deployment:

FIGURE 4.2 The 3-m long 'Wave Glider' by Liquid Robotics (a Boeing company) uses solar and wave energy.

Notes: *WaveGlider: Persistent Ocean Monitoring*, **n.d.**

- **Fixed Installations**: Devices can be mounted on the seabed or attached to underwater structures to provide continuous power to nearby sensor networks (Wang et al., 2024).
- **Mobile Platforms**: Integrating energy harvesters into AUVs or gliders can extend mission duration by recharging on-board systems during operation.
- **Distributed Network**: For large-scale environmental monitoring, arrays of low-power harvesters can power sensor networks that communicate data over long distances.

4.7.7.4 Other Considerations

While energy harvesting technologies offer environmental benefits by reducing the need for battery replacements and manual intervention, their deployment must be managed carefully.

Evaluating the potential impact on marine ecosystems is critical. For example, the electromagnetic fields generated by some inductive charging systems or the physical presence of turbines may affect local fauna.

Deployment must adhere to national and international regulations governing underwater installations, particularly in protected or sensitive marine areas.

4.8 CONCLUSION

AUV energy and propulsion systems are critical, interconnected areas of technology that dictate vehicle capabilities, particularly endurance, range, speed, and manoeuvrability. While significant progress has been made, notably in battery energy density (Lithium-ion, Lithium-polymer) and the development of fuel/semi-fuel cells for longer missions, energy storage remains a primary limitation.

Extending mission duration and distance remains a primary challenge, for which current battery technology is a key constraint. There is a constant need to reduce the size, weight, and power consumption of systems while managing heat, as these factors impede mobility and increase costs. Larger payloads, more powerful processors, and higher speed propulsion all contribute to increased requirements of vehicle Size, Weight, Power, and Cooling (SWaP-C).

The design of these systems involves complex trade-offs related to SWaP-C, efficiency, cost, and reliability, heavily influenced by the intended mission profile and environmental conditions. Continued research and development in these areas, particularly in achieving higher energy densities, improving overall system efficiency, and developing reliable, adaptable propulsion and energy management, are essential for expanding the utility and operational envelopes of future AUVs.

NOTE

1. https://theliquidgrid.com/marine-clean-tech-briefs/ocean-energy-storage/

REFERENCES

AUV ROV Sensors. (n.d.). Sea-Bird Scientific. Retrieved May 2025, from https://www.seabird.com/auv-rov-sensors/family

Bluefin Robotics Unmanned Underwater Vehicles. (n.d.). General Dynamics Mission Systems. Retrieved May 2025, from https://gdmissionsystems.com/underwater-vehicles/bluefin-robotics

Bradley, A. M., Feezor, M. D., Singh, H., & Sorrell, F. Y. (2001). Power Systems for Autonomous Underwater Vehicles. *IEEE Journal of Oceanic Engineering*, *26*(4), 526–538.

Dell, R. M., & Rand, D. A. (2001). *Understanding Batteries*. Royal Society of Chemistry. https://doi.org/10.1039/9781847551065

Griffiths, G. (Ed.). (2003). *Technology and Applications of Autonomous Underwater Vehicles*. Taylor & Francis. https://doi.org/10.1201/9780203023249

Hasan, H. (2017). *Underwater Robotics: Science, Design and Fabrication*. Malaysia: UTM Press: Malaysia.

Liu, J., Yu, F., He, B., & Soares, C. G. (2024, April). A Review of Underwater Docking and Charging Technology for Autonomous Vehicles. *Ocean Engineering*, *297*(1). Retrieved May 2025, from https://www.sciencedirect.com/science/article/abs/pii/S0029801824004918

Massaglia, G. &. (2021). Integration of Portable Sedimentary Microbial Fuel Cells in Autonomous Underwater Vehicles. *Energies*, *14*. Retrieved May 2025, from https://www.researchgate.net/publication/353525521_Integration_of_Portable_Sedimentary_Microbial_Fuel_Cells_in_Autonomous_Underwater_Vehicles

Mohsan, S. A., Khan, M. A., Mazinani, A., Alsharif, M. H., & Cho, H.-S. (2022, September). Enabling Underwater Wireless Power Transfer towards Sixth Generation (6G) Wireless Networks: Opportunities, Recent Advances, and Technical Challenges. *Journal of Marine Science and Engineering*, *10*(9). https://doi.org/10.3390/jmse10091282

Moore, S., Bohm, H., & Jensen, V. (2010). *Underwater Robotics: Science, Design and Fabrication*. Monterey, CA: Marine Advanced Technology Education Center.

Navy SBIR. (2015). Advanced Hybrid Energy System for Wet and Dry Submersibles. https://www.navysbir.com/success/docs/Lynntech_R1.pdf

Ocean Energy Storage. (n.d.). Theliquidgrid.com. Retrieved from https://theliquidgrid.com/marine-clean-tech-briefs/ocean-energy-storage/

Owens, K. (2022, April 13). *DARPA's Power Supply Technology for Undersea Navy Drones*. Warrior Maven-Center for Military Modernization. Retrieved January 2025, from https://warriormaven.com/news/sea/darpa-finds-new-power-supply-for-undersea-navy-drones

Roberts, G. N., & Sutton, R. (Eds.). (2013). *Further Advances in Unmanned Marine Vehicles*. The Institution of Engineering and Technology. https://doi.org/10.1049/PBCE002E

Rutherford, K. T. (2008). *Autonomous Underwater Vehicle Design Considering Energy Source Selection and Hydrodynamics*. University of Southampton.

Teeneti, C., Truscott, T., Beal, D., & Pantic, Z. (2021, January). Review of Wireless Charging Systems for Autonomous Underwater Vehicles. *IEEE Journal of Oceanic Engineering*, *46*(1), 68–87. https://doi.org/10.1109/JOE.2019.2953015

Wang, H, W. Z.-K. (2022, September). Underwater Compressed Gas Energy Storage (UWCGES): Current Status, Challenges, and Future Perspectives. *Applied Sciences*, *12*(18). https://doi.org/10.3390/app12189361

Wang, D., Zhang, J., Cui, S., Bie, Z., Chen, F., & Zhu, C. (2024, January). The State-of-the-arts of Underwater Wireless Power Transfer: A Comprehensive Review and New Perspectives. *Renewable and Sustainable Energy Reviews*, *189*(A). https://doi.org/10.1016/j.rser.2023.113910

WaveGlider: Persistent Ocean Monitoring. (n.d.). Liquid Robotics. Retrieved May 2025, from https://www.liquid-robotics.com/wave-glider/how-it-works/

Yang, Y., Li, L., Yang, J., Tian, H., & Shu, G. (2026). Integrated Design for Underwater Thermoelectric Generator (TEG) Based on Fully Coupled Multiphysics Fields Analysis Method. *Applied Energy*, *407*. https://doi.org/10.1016/j.apenergy.2025.127317

5 UUVs and Sensors

5.1 REQUIREMENTS

Sensors are not only the 'eyes and ears' of an AUV but also serve as its scientific instruments, navigation aids, and communication nodes. Steady advances in sensor miniaturisation, energy efficiency, and intelligent data processing have significantly enhanced the capabilities of AUVs (RAND Corporation, 2019).

AUVs and ROVs carry a diverse range of sensors and payloads designed to meet mission-specific objectives (*AUV ROV Sensors*, n.d.). The sensors and payloads required for the mission objectives should be the starting point of the vehicle design. These sensor systems can be grouped broadly into categories that are described further (Moore, Bohm, & Jensen, 2010).

5.1.1 Environmental and Oceanographic Sensors

- **Conductivity–Temperature–Depth (CTD) Sensors**: CTD sensors are among the most common payloads on AUVs (Figure 5.1). They measure the electrical conductivity, temperature, and pressure (depth) of the surrounding water. Data from CTDs are used to calculate salinity and density, both of which are critical for understanding oceanographic processes. Advances in miniaturisation now allow CTDs to be integrated into compact packages with lower power consumption while maintaining high accuracy (*AUV ROV Sensors*, n.d.).[1]

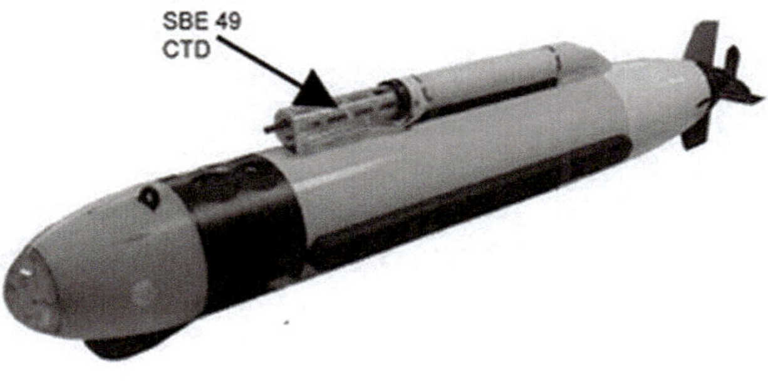

FIGURE 5.1 CTD sensor integrated with an AUV.

Notes: **https://misclab.umeoce.maine.edu/ftp/instruments/CTD%2019+/seabird%20web-site%20disc%20Dec% 202016 /website/application notes/AN93.htm**

- **Acoustic Doppler Current Profiler (ADCP)**: An ADCP measures water current velocities at various depths. It sends sound pulses into the water and then measures the Doppler shift in echoes scattered back by particles suspended in the water. The frequency shift tells the speed and direction of water currents at different depths (i.e., a water column profile).
- **Fluorometers and Turbidity Sensors**: Fluorometers detect chlorophyll concentrations and other fluorescent compounds, providing insights into biological activity and primary productivity. Turbidity sensors measure the clarity of water, which can be used to infer suspended sediment concentrations or detect pollution events (Freitag et al., 2005).
- **Dissolved Oxygen and pH Sensors**: These sensors provide key chemical data on water quality. Dissolved oxygen sensors help assess the habitability of marine environments, while pH sensors are essential for monitoring acidification trends due to climate change (Fletcher, 2001).

5.2 ACOUSTIC SENSORS

Sonars are vital for navigation, mapping, and object detection (Dhanak & Xiros, 2016). Their main types used on UUVs are described further.

5.2.1 Side-Scan Sonar (SSS)

Side-scan sonar systems produce images of the seafloor by measuring the intensity of reflected acoustic signals. They are particularly useful for detecting objects and anomalies on the seafloor. They are useful for imaging the seafloor and detecting objects such as shipwrecks or mines (Figure 5.2 and 5.3) (Moore, Bohm, & Jensen, 2010).

FIGURE 5.2 Edgetech side-scan sonar (black horizontal strip on mid-section) fitted on REMUS100 AUV.

Notes: **https://www.researchgate.net/figure/AUV-REMUS100-with-Edgetech-side-scan-sonar_fig 3_271190867**

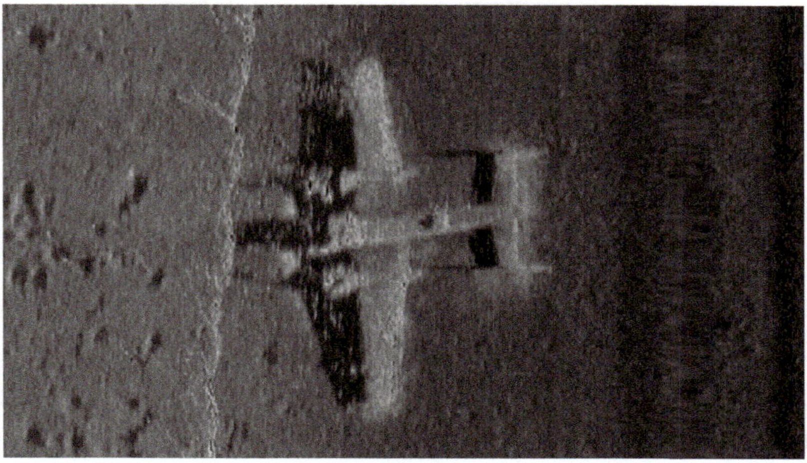

FIGURE 5.3 High resolution side-scan sonar image of a WWII B-25 aircraft discovered on the seabed in 2017 in Papua New Guinea.

Source: **NOAA.**
Notes: **https://oceanexplorer.noaa.gov/technology/sonar/side-scan.html**

Side scan is less expensive to run than, for example, an ROV with a high-definition camera (*Side Scan Sonar*, n.d.). The resulting 'picture' from side-scan sonar data is made up of dark and light areas. Hard objects protruding from the bottom send a strong echo and create a dark image. Shadows and soft areas, such as mud and sand, send weaker echoes and create light areas. These dark and light images help create accurate maps of the seafloor (*Side Scan Sonar*, n.d.).[2] A lower frequency (50–100 kHz) can cover large swathes of the seafloor at low image resolution. Higher-frequency pulses (500 kHz to 1 MHz) record smaller areas but in much greater detail.

5.2.2 Synthetic-Aperture Sonar (SAS)

The SAS provides higher-resolution images by processing multiple pings over time to synthesise a large aperture. It can achieve resolutions down to a few centimetres. A SAS beam is like a 'funnel' that overlaps with itself multiple times to achieve highly detailed images (Figure 5.4). As a result, synthetic aperture sonar can map a site at 30 times the resolution of traditional side scan sonar (*Synthetic Aperture Sonar*, n.d.).[3]

SAS can be used for imaging cultural heritage sites like shipwrecks, classifying habitats or biological organisms, and characterising seafloor sediment makeup (Figure 5.5).

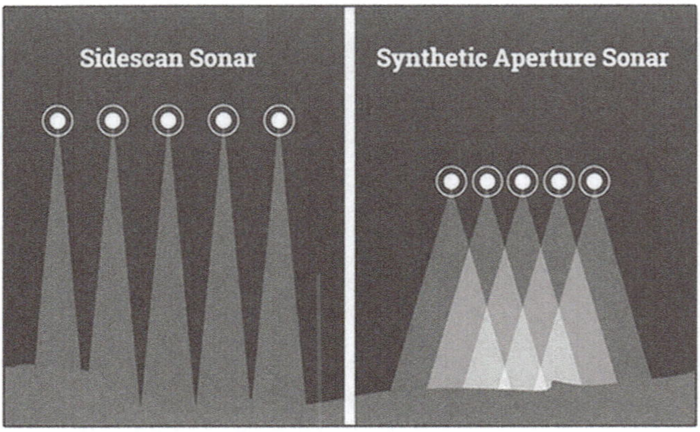

FIGURE 5.4 Comparison of the principle of operation of side-scan sonar and synthetic aperture sonar.

Source: **NOAA.**
Notes: **https://oceanexplorer.noaa.gov/technology/sonar/side-scan.html**

FIGURE 5.5 Kraken Robotics SAS installed on REMUS AUV.

Notes: **https://sea-technology.com/tag/synthetic-aperture-sonar**

5.2.3 Multibeam Echo Sounder (MBES)

This device emits sound waves in multiple beams to map large swaths of the seafloor. It generates high-resolution, detailed bathymetric maps with resolutions typically ranging from 0.5 m to 2 m, depending on frequency and altitude (i.e., height above seabed).

A multibeam sonar uses multiple transducers (in an array) to send and receive sound pulses that map the seafloor or detect other objects. The array sends out multiple, simultaneous sonar beams (or sound waves) at once in a fan-shaped pattern. This covers the space both directly under the vessel and out to each side. Multibeam collects two types of data: seafloor depth and backscatter.

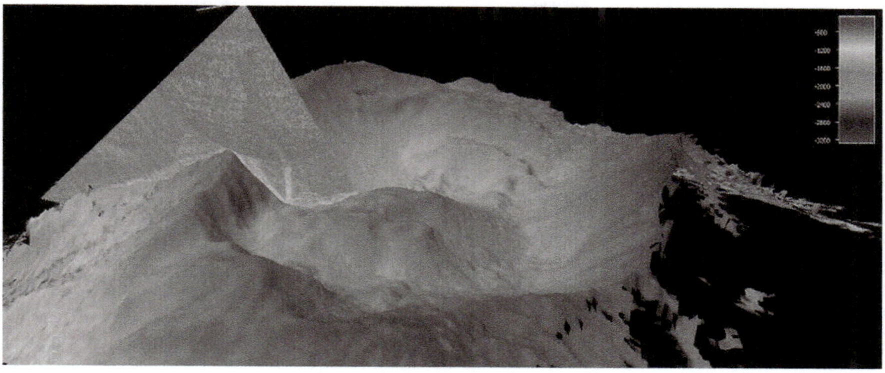

FIGURE 5.6 A plume of bubbles (see the bottom right of the triangle) rising from a seafloor mount, using a single ping of multibeam sonar data.

Source: **NOAA.**
Notes: **https://oceanexplorer.noaa.gov/technology/sonar/multibeam.html**

The seafloor depth, or bathymetry, is computed by measuring the time it takes for the sound to leave the array, hit the seafloor, and return to the array. Backscatter is a measurement of the intensity of the sound echo that reflects back to the multibeam array. These data are processed to create colourful two- or three-dimensional bathymetric maps to visualise the seafloor. In the bathymetric map example shown in Figure 5.6, the warmer colours correspond to shallower areas, while the cooler colours indicate deeper areas (*Multibeam Sonar*, n.d.).[4]

FIGURE 5.7 Multibeam Echosounder (MBES) Surveyor 240-16, weighing 790 grams, size 190 x 90 mm.

Source: **Blue Robotics.**
Notes: **https://bluerobotics.com/store/sonars/echosounders/cerulean-surveyor-240-16-multibeam-echosounder/**

FIGURE 5.8 Multibeam Echosounder (MBES) HydroBeam M4, weighing 5.9 kg, size 230(Dia) × 175(H) mm.

Source: **Satlab.**
Notes: **https://www.satlab.com.se/wp-content/uploads/2025/04/HydroBeam-M4-Brochure-EN-20250422s.pdf**

Multibeam Echosounders find extensive use in Hydrographic Surveying, Offshore Oil and Gas Exploration, Oceanography, Underwater Archaeology, Fisheries Management, Underwater Construction, and Environmental Monitoring. Examples are shown in Figures 5.7 and 5.8.

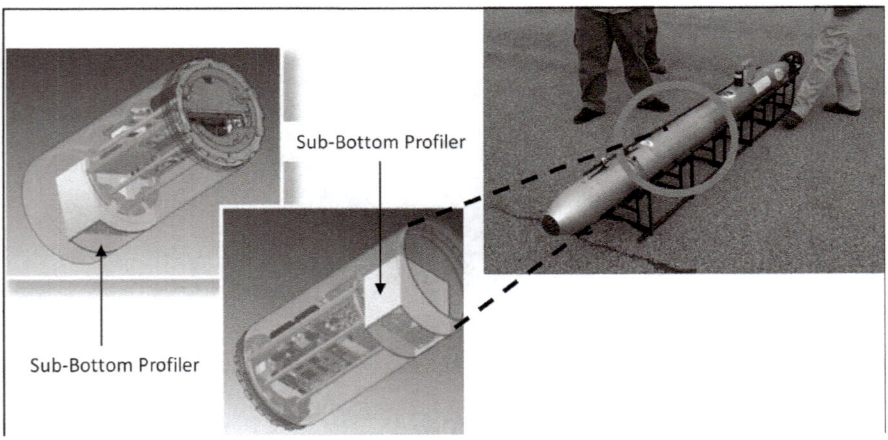

FIGURE 5.9 Teledyne Benthos III Chirp Sub-bottom Profiler fitted inside a Gavia Scientific AUV.

Notes: **https://www.oceanbytes.org/2011/04/01/sub-bottom-profiling-using-an-auv/**

5.2.4 Sub-bottom Profiler

A sub-bottom profiler is a type of sonar system. It is a geophysical survey tool that uses sound to map beneath the seafloor (Figure 5.9).

A sub-bottom profiler emits low-frequency sound pulses that are directed vertically at the seafloor via a transducer. When some of these pulses hit the substrate, they may penetrate the initial layers of the seafloor before reaching the horizons of various sediment types. How far the pulses penetrate, and how long they take to bounce back, will vary based on the makeup of the substrate. By calculating the time taken for a sound pulse to return to the ship, the sediment and layer composition, thickness, and other characteristics are determined.

It is thus possible to generate cross sections or vertical slices depicting sediments below the seafloor. Through interpreting these cross sections, hydrographers can identify features such as volcanic ridges, underwater landslides, ancient river beds, buried objects, and other geological features (*Sub-Bottom Profiler*, 2023).

5.3 OPTICAL SENSORS

- **Cameras and Imaging Systems**: High-definition digital cameras (including low-light and hyperspectral cameras) are increasingly being integrated on AUVs for visual imaging of marine habitats, infrastructure inspection, and object recognition. Advances in CMOS sensor technology and low-light sensitivity have enabled higher-resolution imaging with lower power consumption (Caccia, 2006).
- **LiDAR and Laser Scanning**: Although less common due to limited range in turbid water, underwater LiDAR systems are emerging to provide high-resolution 3D mapping of seafloor structures and complex underwater objects (Griffiths, 2003).

5.4 CHEMICAL AND BIOLOGICAL SENSORS

- **Chemical Analysers**: These sensors can detect dissolved chemicals, nutrients (e.g., nitrate, phosphate), and contaminants. They are often miniaturised for AUV integration, enabling real-time water quality monitoring.
- **Biosensors**: Biosensors, such as those that measure chlorophyll fluorescence or detect microbial activity, help in ecological studies and can be used to monitor harmful algal blooms.
- **Radiometric Sensors**: Radiometers measure light intensity and spectral distribution in underwater environments. They are critical in studying the optical properties of water and in calibrating imaging systems.

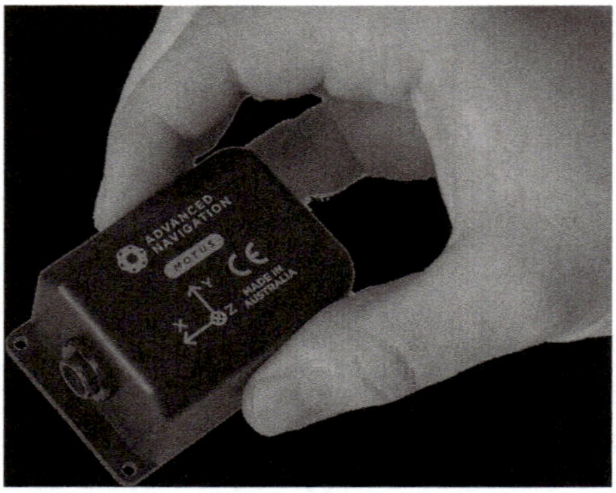

FIGURE 5.10 Motus MEMS IMU weighing 26 g.

Notes: **https://www.unmannedsystemstechnology.com/feature/selecting-an-inertial-measurement-unit-imu-for-uav-applications/**

5.5 NAVIGATION AND POSITIONING SENSORS

5.5.1 Inertial Measurement Units (IMUs)

Inertial navigation systems use accelerometers and gyroscopes (often integrated into an Inertial Measurement Unit, or IMU) to compute the vehicle's position by integrating the measured acceleration and angular rates. IMUs combine accelerometers, gyroscopes, and sometimes magnetometers to provide data on the AUV's orientation and movement (Kongsberg Discovery, 2024). They are essential for dead-reckoning navigation when GPS is unavailable underwater (Tiano & Ferri, 2018).

Even a miniature ultra-high accuracy MEMS IMU (Figure 5.10) can provide inertial performance exceeding that of even some FOG IMUs. It features a bias stability of 0.4°/hr

5.5.2 Acoustic Positioning Systems

Systems such as USBL (ultra-short baseline) and LBL (long baseline) allow the AUV to determine its position relative to known acoustic beacons, thereby supplementing inertial navigation (Stojanovic, 2007). These can also be categorised under 'acoustic' sensors. They aid in collaborative missions among multiple AUVs (Hasan, 2017).

5.5.2.1 Ultra-Short Baseline (USBL)

The USBL includes two modes: the transponder mode and the responder mode. In the transponder mode, an interrogation signal is sent from the surface transducer array to the seabed transponder (L3Harris, 2022). After this interrogation signal is received, an acknowledge signal is transmitted from the transponder. The time delay between the transmission of the interrogation signal and the reception of the acknowledge signal is used for calculating the range (Zhang, Liu, & Liu, 2018).

In the responder mode, a pair of high-precision clocks is used to trigger the responder on the submersible and USBL surface processing unit on the mother ship synchronously. The time delay between the synchronous pulse triggering moment and the reception of the acknowledge signal is used for calculating the slant range. The sound speed profile is needed for calculating the range from the time delay. Thus, an accurate sound speed profile guarantees accurate positioning. In addition to the range, the azimuth is needed for positioning and can be computed by analysing the phase of the received signals. Combining the GPS and USBL, the exact location of the submersible can be obtained (Zhang, Liu, & Liu, 2018).

5.5.2.2 Long Baseline (LBL)

The LBL system consists of two parts: a transducer and processing unit installed in the manned submersible, and a set of beacons at the bottom of the ocean (Zhang, Liu, & Liu, 2018). The LBL acoustic positioning converts the travel times to ranges utilising three or more widely spaced (approximately 4 km) stationary beacons for calculating the location of a moving vehicle in two or three dimensions. The beacons are typically stationary (fixed baseline) and moored to the seafloor with tethers. Alternatively, they can be held at fixed relative locations on a moving platform, such as a ship (moving baseline). At least three beacons (known positions) are required to form a specific geometry. The beacon locations must be determined during an initial offline survey (L3Harris, 2022).

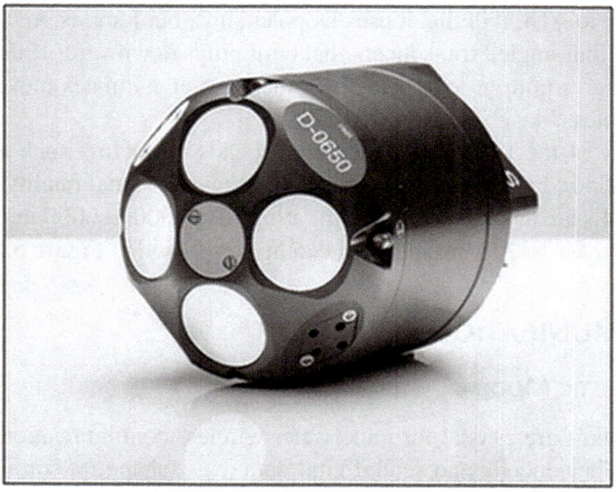

FIGURE 5.11 Doppler velocity log.

Source: Nortek.
Notes: **https://www.nortekgroup.com/products/dvl-1000-300m**

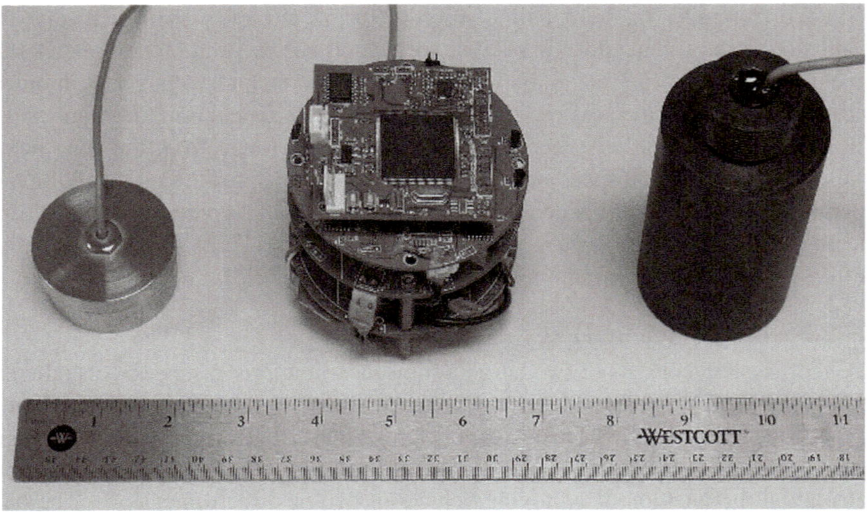

FIGURE 5.12 Acoustic modem (middle) with directional (left) and omnidirectional transducer (right).

Notes: **https://www.researchgate.net/figure/Acoustic-modem-middle-with-directional-left-and-omnidirectional-transducer-right_fig 1_329617107**

5.5.3 Doppler Velocity Log (DVL)

DVLs measure the vehicle's velocity relative to the seafloor (or water column) by transmitting acoustic pulses and measuring the Doppler shift of the returned signal (*DVL1000*, n.d.).

It is similar to ADCP in that it uses Doppler shift, but focuses on vehicle motion. It usually has four angled transducers that emit pings downward. If the bottom is in range, it measures bottom-lock velocity. Otherwise, it estimates movement relative to water particles (like an inverted ADCP).

The quality of the DVL measurements depends on factors such as the altitude above the seafloor, bottom characteristics, and acoustic signal quality (Hegrenæs et al., 2016). DVLs are available with depth ratings from 300 to 6000 m, with a bottom track from over 0.1 to 200 m range. An example is shown in Figure 5.11.

5.6 COMMUNICATION EQUIPMENT

5.6.1 Acoustic Modem

Acoustic modems are critical for underwater wireless communication (Inzartsev & Pavi, 2009). They encode and send digital data (e.g., telemetry, commands, or sensor readings) in the form of acoustic signals. Their receiver decodes the signals into usable information. Typical characteristics are elaborated in the chapter on underwater communication (Figure 5.12).

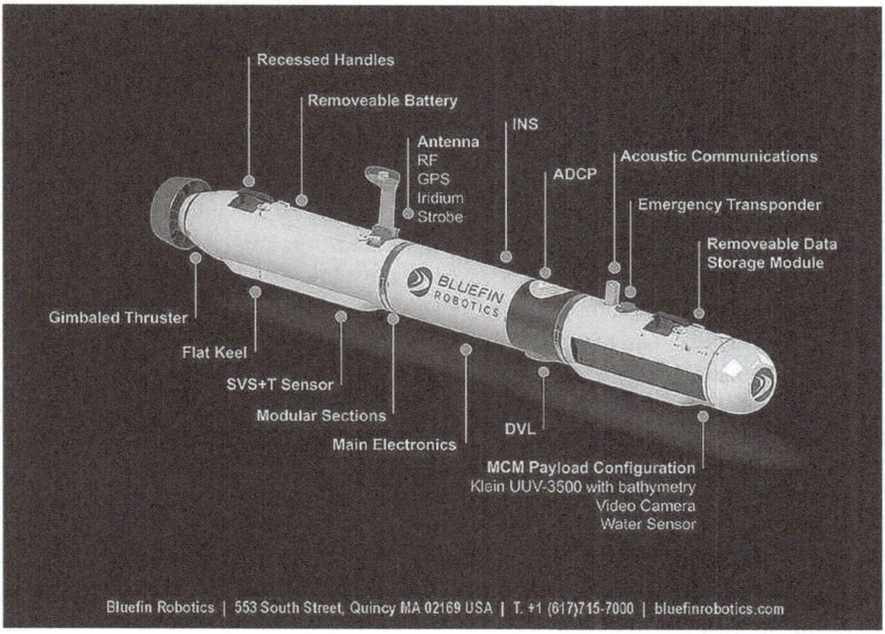

FIGURE 5.13 Arrangement of sensors and other components on Bluefin-9M AUV.

Source: **General Dynamics.**
Notes: **https://gdmissionsystems.com/discontinued-products/bluefin-9-m-autonomous-underwater-vehicle**

5.6.2 GPS Antenna

For UUVs, GPS antennas are crucial for surface navigation and positioning when the vehicle surfaces. They are often used in conjunction with other underwater positioning systems like acoustic modems or inertial navigation systems (INS). These antennas need to be rugged, waterproof, depth-rated, and capable of operating in harsh marine environments (Alexandris, Papageorgas, & Piromalis, 2024).

5.6.3 RF Antenna

AUVs and UUVs often utilise UHF (225–400 MHz) for LOS communication and potentially L-band or S-band for SATCOM. Due to the limited space within AUVs and UUVs, compact antennas are essential.

The antenna must withstand the pressure, salinity, and potential impacts associated with underwater operations. UHF antennas are commonly used for LOS communication between the AUV/UUV and a surface vessel or base station. L-band or S-band antennas enable communication with satellites, facilitating long-range communication and data transfer. Antennas that support mesh networking can facilitate communication between multiple AUVs or UUVs, enabling collaborative operations

(Zhou, 2021). Some antennas are designed to receive GPS signals while the AUV/UUV is submerged, allowing for accurate positioning (Figure 5.13).

5.7 SENSOR SELECTION AND INTEGRATION

Modern AUVs are increasingly designed with modular payload bays that can accommodate different combinations of sensors based on mission requirements. For instance, an AUV deployed for mine countermeasures might integrate synthetic-aperture sonar with magnetometers and high-resolution video cameras, while one used for oceanographic sampling might carry a suite of CTD sensors, fluorometers, and water samplers. The flexibility in payload integration is a key advantage, allowing the same platform to be reconfigured for diverse missions (Allen et al., 2002).

Sensor selection for UUVs must balance performance requirements, environmental conditions, integration constraints, and mission objectives (Carder et al., 2001). These factors are examined further.

5.7.1 Mission Requirements

The primary driver of sensor selection is the mission profile, which could be any (or all) of the following:

- **Scientific Research**: High-precision measurements, multi-parameter data collection, and long-duration deployments demand sensors with high resolution and stability.
- **Military and Defence**: Robustness, stealth, and the ability to detect and classify objects (e.g., mines, submarines) are paramount.
- **Industrial Inspections**: Sensors used in offshore pipeline monitoring or infrastructure inspections need to be rugged, reliable, and capable of delivering real-time data.

Mission requirements determine the type of data needed (e.g., acoustic imaging or optical imaging), the resolution of that data, and the acceptable error margins (Chocron et al., 2017). In addition, mission duration, area coverage, and the need for autonomous decision-making influence sensor selection (Griffiths, 2003).

5.7.2 Environmental Conditions

Sensors must be selected based on the environment in which the AUV will operate. Some of the important factors are listed below.

- **Depth and Pressure**: Deep-sea missions require sensors that can withstand high hydrostatic pressures. Pressure-tolerant or pressure-compensated designs are essential.

- **Temperature**: Extreme cold or warm conditions can affect sensor performance. Components must be able to operate over the anticipated temperature range.
- **Water Clarity and Turbidity**: Optical sensors are highly sensitive to water clarity. In turbid environments, sensors with enhanced low-light performance or alternative imaging modalities (e.g., sonar) may be necessary.
- **Salinity and Corrosiveness**: Saltwater is highly corrosive, so sensor materials and housings must be selected for durability and resistance to biofouling.

5.7.3 Power Consumption and Energy Budget

AUVs are energy-constrained platforms. Sensor power requirements must be balanced against the overall energy budget of the vehicle (Defense Science Board, 2015).

- **Low-Power Design**: Many sensors are designed to operate at low power (ranging from a few milliwatts to a few watts). However, high-resolution imaging systems or active sonar modules may require significantly more power (from 5 W to tens of watts).
- **Duty Cycle and Operational Modes**: Sensors that do not need to operate continuously can be scheduled to power on only when needed, reducing overall energy consumption.
- **Integration with Energy-Harvesting or In-Situ Charging**: Emerging technologies such as underwater wireless power transfer or energy harvesting (e.g., from wave energy) can supplement battery power, particularly for long-duration missions.

5.7.4 Weight, Size, and Space Constraints

Due to the limited payload capacity and space within an AUV, sensor size and weight are critical considerations.

Advances in MEMS (micro-electro-mechanical systems) and CMOS sensor technology have led to smaller, lighter sensors that still maintain high performance. Miniature size is an advantage.

Sensor packages must be designed to fit within the available payload bays while leaving room for other critical components (e.g., batteries, communication systems). Thus, integration into the vehicle must be compact.

A modular approach allows different sensor suites to be swapped in and out without major modifications to the AUV's overall architecture. Many modern AUVs, such as the REMUS series, incorporate plug-and-play payload modules that optimise available space (*REMUS UUVs*, n.d.).[5]

5.7.5 Data Bandwidth and Processing Requirements

Sensors that generate large volumes of data (such as high-definition video or synthetic-aperture sonar) impose significant demands on on-board storage and processing capabilities (Choi & Yuh, 2016).

- **Real-Time Processing**: In missions where immediate data interpretation is crucial (e.g., object detection in mine countermeasures), sensors must be paired with powerful on-board processors that can analyse data in real time.
- **Compression and Storage**: High-data-rate sensors require efficient compression algorithms to store and transmit data without overwhelming the AUV's communication links.
- **Communication Bandwidth**: Underwater communication channels (typically acoustic) have limited bandwidth. Sensors must be selected and configured so that data output is manageable given these constraints.

5.7.6 Reliability, Robustness, and Maintenance

The harsh underwater environment necessitates sensors that are robust and require minimal maintenance, and the relevant factors are listed.

- **Corrosion Resistance**: Materials and enclosures must resist corrosion from salt water.
- **Biofouling Prevention**: Sensors deployed for long durations may suffer from biofouling, which can degrade performance. Anti-fouling coatings and mechanical cleaning systems are sometimes employed.
- **Long-Term Stability**: Sensors must maintain calibration and accuracy over extended missions without the need for frequent recalibration.
- **Ease of Replacement and Repair**: In modular systems, sensors should be easily replaceable to reduce downtime during maintenance.

5.7.7 Cost and Economic Considerations

Finally, the cost of sensors can be a limiting factor, particularly for research projects or large-scale deployment.

- **Cost-Performance Trade-Off**: While high-end sensors may offer superior accuracy and resolution, they can significantly increase project costs. Engineers must balance performance requirements with available budgets.
- **Scalability**: For applications that require large sensor networks, such as the Subsea Internet of Things (SIoT) or Underwater Wireless Sensor Networks (UWSNs), cost-effective, scalable sensor solutions are preferred (Gupta & Singh, 2017).

5.8 ENERGY, POWER, WEIGHT, AND SPACE

Understanding the typical specifications of AUV sensor systems is key to designing an effective payload suite. Although figures vary based on sensor type and manufacturer, the following guidelines provide a general overview.

5.8.1 Energy and Power Requirements

- **Low-Power Sensors**: Many environmental sensors (e.g., CTDs, temperature, conductivity) operate on the order of milliwatts (typically 50 mW to 500 mW). Their design emphasises energy efficiency to allow long-term deployment. They are often operated intermittently to conserve energy. Chemical/biological sensors typically range from 50 mW to 1 W, depending on the sensor type and duty cycle.
- **Active Sonar Systems**: Sonar sensors, particularly those with active transmission (such as synthetic-aperture sonar or multibeam sonars), may require power levels ranging from 5 W to 50 W, depending on pulse power, beamforming requirements, and resolution. Active Sonar (e.g., SAS, Multibeam) can require between 5 W and 50 W when transmitting pulses. Passive Sonar (e.g., hydrophones) usually consumes less power (approximately 1–5 W) (Aoki, 2008).
- **Optical Imaging Systems**: High-definition video cameras or hyperspectral imagers can range from 1–5 W for low-resolution sensors up to 10–20 W for high-end, high-resolution systems, particularly when processing is done on board (Balasuriya & Ura, 2002).
- **Data Processing and On-board Computing**: Depending on the complexity of the processing (e.g., real-time object recognition, adaptive control algorithms), on-board computers may require anywhere from 5 W to 30 W.
- **Total Payload Power**: In many integrated AUV payload suites, the total power requirement may range from 10 W to 100 W. For missions requiring high-resolution imaging or high-data-rate acoustic systems, power consumption may exceed 100 W (Bachmayer et al., 2008). Since AUVs are battery-operated with finite energy capacity, sensor duty cycles are managed carefully to conserve power.

5.8.2 Weight and Space Considerations

Advances in sensor miniaturisation have enabled many sensors to be produced in compact packages. However, sonar and communication sensors may occupy significant space relative to vehicle size and may also govern the vehicle configuration. Salient considerations and typical values are given here (Tijjani & Chemori, 2021).

- **Imaging Sensors**: High-definition cameras used on AUVs generally weigh between 0.5 kg and 2 kg. When combined with additional optics and housing for underwater use, a complete imaging system might range from 1 kg to 3 kg. These systems are designed to be compact, often occupying less than 2 L.

- **Sonar Systems**: Acoustic transducers and sonar modules vary in size. Small single-beam sonars can weigh a few hundred grams, while multi-beam or synthetic-aperture sonar arrays may weigh several kilograms and require larger physical housings (Kongsberg Maritime, 2023). They often require a dedicated mounting structure, with typical volumes in the range of 1–5 L.
- **Chemical and Biological Sensors**: Many of these sensors are designed using microfluidic techniques and MEMS technology; they can be very lightweight (often less than 500 g) and compact (typically a few cubic decimetres). A modern CTD sensor may weigh less than 1 kg and occupy a volume of 0.5–1 L.
- **Payload Bay Constraints**: AUV platforms often have strict space and weight limits. For example, smaller AUVs (such as the REMUS 100) may have a payload capacity of only a few kilograms, whereas larger AUVs (like the REMUS 6000) can carry payloads of tens of kilograms. Modular design strategies help maximise the effective use of space by stacking or integrating sensor packages (Figure 5.14) (Billon-Coat, 2018).

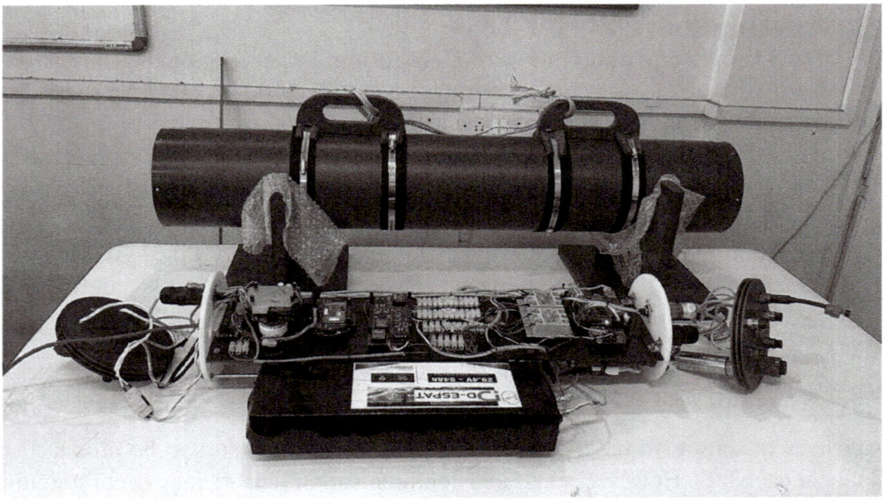

FIGURE 5.14 AUV Jaya components before integration.

Source: **Author/NIO Goa.**

- **Navigation Sensors (IMUs, DVLs, USBL modules)**: Weight can vary from 0.5 kg (for simple IMUs) to 2–3 kg for integrated navigation packages. These are typically designed for a minimal footprint and can be integrated into the vehicle's hull or internal frame.
- **Integration**: Sensors must be mounted securely and oriented correctly. The integration may involve custom-designed mounting brackets, waterproof housings, and shock absorbers. Designers strive for a balance between robust sensor placement and minimal interference with the hydrodynamic profile of the vehicle (Rutherford, 2008). COTS sensors that need to be exposed to seawater usually come in pressure-resistant housing and need to be interfaced and calibrated after integration on the vehicle.

5.9 TYPICAL PAYLOAD SPECIFICATIONS

Consider a typical mid-range AUV used for environmental and geophysical surveys. A representative sensor payload might include the following:

- **CTD Sensor**: Weighing ~0.8 kg, occupying 1 L, with a power consumption of 200 mW.
- **Multibeam Sonar**: A compact multibeam sonar module weighing ~3 kg, requiring 30 W during operation.
- **High-Definition Camera System**: A digital imaging package weighing ~1.5 kg, consuming 5 W continuously.
- **Doppler Velocity Log (DVL)**: Weighing ~2 kg, with power consumption around 10 W.
- **Additional Chemical Sensors (e.g., dissolved oxygen, pH, nitrate)**: Combined weight of ~1 kg, power consumption of ~500 mW each (with duty cycling, average consumption might be lower).

For this configuration, the total payload might weigh approximately 8–10 kg and require an average power budget of roughly 50–60 W when all sensors are operating continuously. In practice, many sensors do not run continuously (they may be duty cycled), so the overall energy consumption is reduced (Roberts & Sutton, 2013). Integration constraints, including available payload bay volume and mass limits, drive design trade-offs between sensor quality, size, and energy use.

These typical specifications serve as a guideline for AUV payload designers, who must often customise sensor configurations based on the specific vehicle size and mission parameters. In many cases, designers rely on modular payload bays that allow the same AUV platform to be configured with different sensor suites.

5.9.1 Examples of UUV Sensor Configurations

Several AUV platforms can serve as case studies of how sensor payloads are integrated and utilised (Wadoo & Kachroo, 2011). A few widely used AUV designs are discussed here.

TABLE 5.1
Typical AUVs and Their Sensors/Payloads

AUV/Vehicle	Sensor/Payload Type	Measured Variable/Function	Characteristics
P-SURO (hovering-type)	Colour camera	Underwater vision	Mounted at vehicle nose
	Range sonar (3 units)	Forward, backwards, and downward ranging	Assists navigation, obstacle avoidance, and SLAM
MARES (small size)	Sea-Bird Electronics 49 FastCAT CTD	Conductivity, temperature, depth	Installed on board
	Omnidirectional acoustic transducer	Long baseline navigation	Supports long baseline navigation
Urashima (deep and long range)	TV camera	Observation, exploration	
	Digital still camera	Observation, exploration	
	CTDO sensor	Conductivity, temperature, depth, dissolved oxygen	
	Side scan sonar (SSS)	Seabed imaging, hazard object detection	Dual frequency (120/410 kHz), max range 500 m
	Multibeam echo sounder (MBE)	Bathymetry observation	Uses 400 kHz, range ~300 m, standalone mode
	Sub bottom profiler (SBP)	Sub-bottom structure and geophysical mapping	Chirp sonar (1–6 kHz), max range ~50 m, 6 receiver channels
HUGIN 3000	Multibeam echo sounder (MBE)	Bathymetry survey	High resolution, depending on the application
	Side scan sonar	Hazard object detection	Depending on the application
	Sub-bottom profiler (SBP)	Sub-bottom structure and geophysical mapping	Depending on the application
	CTD sensor	Conductivity, temperature, depth	Depending on the application
	Magnetometer	Detection of magnetic objects	Depending on the application
	Fishery research echosounder	Used for fishery research	Depending on the application
CentrUMS-ITB (hybrid AUV)	Obstacle avoidance sonar	Obstacle avoidance	Forward looking
	CTD	Conductivity, temperature, depth	Standard sensor payload

(Continued)

TABLE 5.1 (CONTINUED)
Typical AUVs and Their Sensors/Payloads

AUV/Vehicle	Sensor/Payload Type	Measured Variable/Function	Characteristics
JHUROV (ROV example)	SHARP acoustic transponder	3DOF-vehicle position (also navigation)	Precision 0.5 cm, Update rate 10 Hz
	Foxboro/ICT model n. 15	Depth	Precision 2.5 cm, Update rate 20 Hz
Autosub AUV	Side-scan sonar	Seabed imaging, object detection	Standard sensor
	Acoustic velocimeters	Spatial/temporal variations of flow and turbulence	Used in terrain-following missions
	Acoustic air-bubble resonator	Measurement of bubble clouds	Used in combination with other sensors
	Fishery research echosounder	Population density/abundance of fish/krill	Same as used on research vessel
	Spectrophotometric dissolved manganese sensor	Dissolved manganese concentration	Unique adaptation, wet chemistry sensor
	Flow cytometer (Cytobuoy)	Particle size, optical pulse profiles (for plankton analysis)	Autonomous operation, internal memory
	AC9 scattering and absorption meter	Particle content and optical properties of the water column	Integrated with a flow cytometer & a water sampler
Isurus AUV	CTD (OS200 model)	Conductivity, temperature, depth	Environmental monitoring of wastewater discharges
	Altimeter (Imagenex)	Height from the sea bottom	
ICTINEU AUV	MSIS (mechanically scanning imaging sonar)	Imaging horizontal plane, obstacle detection	Configured to scan 360 sector, 50 m range, 0.1 m resolution
Delphin2 AUV	Downwards camera	Seabed imaging, visual detection	
	Forward-looking sonar	Obstacle detection, terrain detection	Single-beam based detection
MIMOSA AUV	Tritech Gemini sonar	Multibeam imaging sonar	Used for sensing other AUVs as opposed to salient obstacles

FIGURE 5.15 Bluefin-21 Macrura UUV deployed in ICEX 2020.

Source: **Dan McDonald, General Dynamics Mission Systems.**
Notes: **https://gdmissionsystems.com/articles/2021/04/26/in-the-news-us-navy-deploys-autonomous-uuvs-under-the-ice**

The specific combination of sensors and payloads varies greatly depending on the AUV model, its size class (man-portable, lightweight, heavyweight, large), and its intended missions. Table 5.1 gives examples of AUVs and their sensors/payloads.

5.9.2 Bluefin

The Bluefin-21 AUV is a well-known platform used in high-profile missions such as the search for Malaysia Airlines Flight 370 (*Bluefin Robotics Unmanned Underwater Vehicles*, n.d.) (Figure 5.15).[6] Its payload typically includes the following:

- **Acoustic Sensors**: An EdgeTech 2200-M sonar, a sub-bottom profiler, and an echo sounder enable detailed mapping of the seafloor.
- **Optical Imaging**: A Prosilica high-resolution camera is used for visual documentation.
- **Navigation**: An inertial navigation system (INS) and an ultra-short baseline (USBL) positioning system provide accurate location data.
- **Energy Considerations**: Bluefin-21 is powered by nine lithium-polymer battery packs (each rated at 1.5 kWh), providing a total energy capacity of approximately 13.5 kWh. This design balances payload weight (typically a few kilograms per sensor module) with the energy required for long-duration missions (*Bluefin Robotics Unmanned Underwater Vehicles*, n.d.).

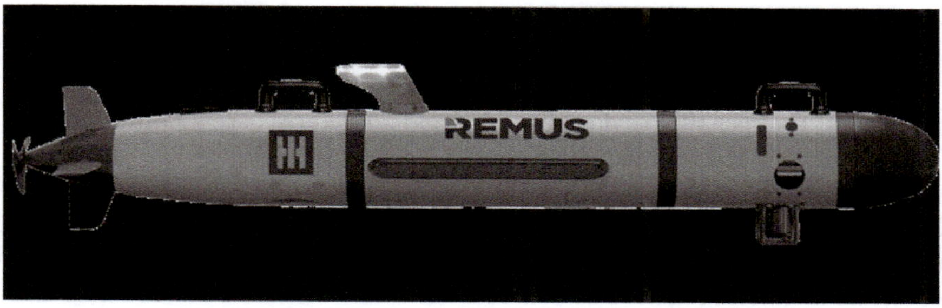

FIGURE 5.16 REMUS UUV.

Source: **HII.**
Notes: **https://hii.com/what-we-do/capabilities/unmanned-systems/remus-uuvs/**

This integrated payload allows Bluefin-21 to conduct high-resolution sonar mapping and detailed visual inspections over extended missions (Bhatt, Viquez, & Schmidt, 2022).

5.9.3 REMUS

The REMUS family of AUVs, developed in collaboration with institutions like Woods Hole Oceanographic Institution (WHOI), is renowned for its modular payload design (*REMUS AUV*, n.d.).[7]

- **REMUS 6000**: This larger vehicle can carry multiple payloads, including multibeam sonars, side-scan sonars, CTDs, and high-definition cameras. Its design emphasises modularity, enabling rapid swapping of sensor packages (*REMUS UUVs*, n.d.).

FIGURE 5.17 Sentry AUV developed by WHOI.

Notes: **https://www.whoi.edu/what-we-do/explore/underwater-vehicles/auvs/sentry/**

- **REMUS 300 and REMUS 100**: These smaller platforms typically integrate essential navigation sensors (IMU, DVL, pressure sensors) along with basic imaging and environmental sensors for scientific surveys (Figure 5.16). Their smaller payload bays demand miniaturised, low-power sensors (Corfield & Hillenbrand, 2003).

The REMUS series illustrates how sensor payloads are tailored to mission requirements—from detailed seabed mapping in the REMUS 6000 to high-frequency environmental monitoring in the REMUS 100.

5.9.4 Sentry

'Sentry' is an AUV designed by the Woods Hole Oceanographic Institution (WHOI) for deep-sea research (Figure 5.17) (*Sentry*, n.d.). Sentry exemplifies the integration of high-precision, deep-sea sensors into an AUV platform designed for extreme conditions (Corfield & Hillenbrand, 2003). Its sensor fit is as follows:

- **Navigation Sensors**: Sentry uses a Doppler Velocity Log (DVL) and inertial navigation systems for precise positioning.
- **Environmental and Imaging Sensors**: It is equipped with CTD sensors, digital cameras, and can carry additional instruments such as magnetometers and chemical analysers.
- **Power and Integration**: Sentry's sensors are integrated into a system designed for deep deployments (up to 6000 m), and its payload modules are optimised for low power consumption to extend mission duration.

5.10 CONCLUSION

The sensors and payloads carried by autonomous underwater vehicles are at the core of their operational capabilities. Today's AUVs integrate a wide variety of sensors, ranging from environmental and acoustic to optical and chemical, in order to perform diverse missions of deep-sea exploration, military surveillance, and environmental monitoring. Sensor selection for AUVs is driven by multiple factors: mission objectives, environmental conditions, power and energy constraints, weight and space limitations, data processing needs, and overall cost (Finn & Scheding, 2010). In many cases, trade-offs must be made to balance performance with efficiency and reliability.

Recent developments in sensor technology, including advances in miniaturisation, smart sensor design, low-power electronics, and hybrid sensor modalities, are opening new possibilities (Martin et al., 2019). Emerging trends such as adaptive, self-calibrating sensors, energy harvesting techniques, and the integration of the Internet of Underwater Things (IoUT) appear promising (Kilfoyle & Baggeroer, 2000).

As sensor technology continues to evolve, the future of underwater sensing will likely feature compact, energy-efficient, and intelligent sensor suites capable of delivering unprecedented levels of data quality. The convergence of sensor technology with advanced on-board processing and better communications should enable real-time data analysis and autonomous decision-making. These improvements will further expand the applications of AUVs in scientific, commercial, and military domains.

NOTES

1. https://www.seabird.com/auv-rov-sensors/family
2. https://oceanexplorer.noaa.gov/technology/sonar/side-scan.html
3. https://oceanexplorer.noaa.gov/technology/sonar/sas.html
4. https://oceanexplorer.noaa.gov/technology/sonar/multibeam.html
5. https://hii.com/what-we-do/capabilities/unmanned-systems/remus-uuvs/
6. https://gdmissionsystems.com/underwater-vehicles/bluefin-robotics
7. https://www.whoi.edu/what-we-do/explore/underwater-vehicles/auvs/remus/

REFERENCES

Alexandris, C., Papageorgas, P., & Piromalis, D. (2024). Positioning Systems for Unmanned Underwater Vehicles: A Comprehensive Review. *Applied Sciences*, *14*(21). https://doi.org/10.3390/app14219671

Allen, B., Stokey, R., Austin, T., Forrester, N., Goldsborough, R., Purcell, M., & Von Alt, C. (2002). REMUS: A Small, Low Cost AUV; System Description, Field Trials and Performance Results. In *Oceans '97. MTS/IEEE Conference Proceedings* (Vol. 2, pp. 994–1000). IEEE. https://doi.org/10.1109/oceans.1997.624126

Aoki, T. (2008). Urashima. In T. Ura (Ed.), *Autonomous Underwater Vehicles, Design and Operation* (2nd ed.). Ura.

AUV ROV Sensors. (n.d.). Sea-Bird Scientific. Retrieved May 2025, from https://www.seabird.com/auv-rov-sensors/family

Bachmayer, R., Antonelli, E., Fossen, T., Indiveri, G., Yuh, J., & Ødegaard, A. (2008). *Springer Handbook of Robotics*. Heidelberg: Springer.

Balasuriya, A., & Ura, T. (2002). Autonomous Target Tracking by Underwater Robots Based on Vision. In *Proceedings of 1998 International Symposium on Underwater Technology* (Vol. 29, pp. 1351–1367). IEEE. https://doi.org/10.1109/ut.1998.670089

Bhatt, E. C., Viquez, O., & Schmidt, H. (2022). Under-ice Acoustic Navigation Using Real-time Model-aided Range Estimation. *Journal of the Acoustical Society of America*, *151*(4), 2656–2671. https://doi.org/10.1121/10.0010260

Billon-Coat, A. (Ed.). (2018). *Marine Robotics and Applications*. Springer International Publishing AG. https://doi.org/10.1007/978-3-319-70724-2

Bluefin Robotics Unmanned Underwater Vehicles. (n.d.). General Dynamics Mission Systems. Retrieved May 2025, from https://gdmissionsystems.com/underwater-vehicles/bluefin-robotics

Caccia, M. (2006). Vision-based Linear Motion Estimation for Unmanned Underwater Vehicles. In *2003 IEEE International Conference on Robotics and Automation (Cat. No.03CH37422). 1* (pp. 977–982). IEEE. https://doi.org/10.1109/robot.2003.1241719

Carder, K., Costello, D., Warrior, H., Langebrake, L., Hou, W., Patten, J., & Kaltenbacher, E. (2001, October). Ocean-science Mission Needs: Real-time AUV Data for Command, Control, and Model Inputs [West Florida Shelf]. *IEEE Journal of Oceanic Engineering*, *26*, 742–751. https://doi.org/10.1109/48.972116

Chocron, O., Chemori, A., Creuze, V., & Lozano, R. (2017). Design and Control of Underwater Vehicles. In L. Jaulin, E. Maître, & J. Yuh (Eds.), *Marine Robotics and Applications* (pp. 121–145). Cham: Springer International Publishing.

Choi, H.-T., & Yuh, J. (2016). Underwater Robots. In B. Siciliano, & O. Khatib (Eds.), *Springer Handbooks* (pp. 595–622). Heidelberg: Springer International Publishing. https://doi.org/10.1007/978-3-319-32552-1_25

Corfield, S., & Hillenbrand, C. (2003). Defence Applications for Unmanned Underwater Vehicles. In G. Griffiths (Ed.), *Technology and Applications of Autonomous Underwater Vehicles* (pp. 161–178). London: CRC Press. https://doi.org/10.1201/9780203522301.ch10

Defense Science Board. (2015). *Next-Generation Unmanned Undersea Systems.* Technical report. Washington, DC: U.S. Department of Defense.

Dhanak, M., & Xiros, N. (2016). *Springer Handbook of Ocean Engineering.* Springer.

DVL1000. (n.d.). Nortek. Retrieved May 2025, from https://www.nortekgroup.com/products/dvl-1000-300m

Finn, A., & Scheding, S. (Eds.). (2010). *Developments and Challenges for Autonomous Unmanned Vehicles: A Compendium.* Berlin Heidelberg: Springer-Verlag. https://doi.org/10.1007/978-3-642-10704-7

Fletcher, B. (2001). Chemical Plume Mapping with an Autonomous Underwater Vehicle. In *MTS/IEEE Oceans 2001. An Ocean Odyssey. Conference Proceedings (IEEE Cat. No.01CH37295).* (Vol. 1, pp. 508–512). Honolulu: Marine Technology Society. https://doi.org/10.1109/oceans.2001.968774

Freitag, L., Johnson, M., Singh, S., Grund, M., Kampf, S., Sherry, J., & Camilli, R. (2005). A Long-range Autonomous Underwater Vehicle for Deep-ocean Exploration at Hydrothermal Vents. *Journal of Field Robotics, 22*, 543–569.

Griffiths, G. (Ed.). (2003). *Technology and Applications of Autonomous Underwater Vehicles.* Taylor & Francis. https://doi.org/10.1201/9780203023249

Gupta, S., & Singh, N. (2017). Underwater Wireless Sensor Networks: A Review of Routing Protocols, Taxonomy, and Future Directions. *The Journal of Supercomputing, 80*, 5163–5196. https://doi.org/10.1007/s11227-023-05646-w

Hasan, H. (2017). *Underwater Robotics: Science, Design and Fabrication.* Malaysia: UTM Press: Malaysia.

Hegrenæs, Ø., Ramstad, A., Pedersen, T., & Velasco, D. (2016). Validation of a New Generation DVL for Underwater Vehicle Navigation. In *IEEE/OES Autonomous Underwater Vehicles (AUV)* (pp. 342–348). Tokyo: IEEE. https://doi.org/10.1109/AUV.2016.7778694

Inzartsev, A., & Pavi, A. (2009, January 1). AUV Application for Inspection of Underwater Communications. In A. V. Inzartsev (Ed.), *Underwater Vehicles* (pp. 371–398). Rijeka: InTech. https://doi.org/10.5772/6704

Kilfoyle, D. B., & Baggeroer, A. B. (2000, January). The State of the Art in Underwater Acoustic Communication. *IEEE Journal of Oceanic Engineering, 25*(1), 4–27. https://doi.org/10.1109/48.820733

Kongsberg Discovery. (2024, September 4). Smashing AUV Records, HUGIN Endurance Completes a Multi-week Fully Autonomous Mission. *Kongsberg.com.* Retrieved May 2025, from https://www.kongsberg.com/newsroom/news-archive/2024/smashing-auv-records-hugin-endurance-completes-a-multi-week-fully-autonomous-mission/

Kongsberg Maritime. (2023). *HiPAP and cNODE Positioning Systems.*

L3Harris. (2022). *USBL Acoustic Positioning Systems.*

Martin, B., Tarraf, D. C., Whitmore, T. C., DeWeese, J., Kenney, C., Schmid, J., & DeLuca, P. (2019). *Advancing Autonomous Systems: An Analysis of Current and Future Technology for Unmanned Maritime Vehicles.* CA: RAND Corporation. https://doi.org/10.7249/RR2664

Moore, S., Bohm, H., & Jensen, V. (2010). *Underwater Robotics: Science, Design and Fabrication.* Monterey, CA: Marine Advanced Technology Education Center.

Multibeam Sonar. (n.d.). NOAA Ocean Exploration. Retrieved January 2025, from https://oceanexplorer.noaa.gov/technology/sonar/multibeam.html

RAND Corporation. (2019). *Advancing Autonomous Systems: An Analysis of Current and Future Technology for Unmanned Maritime Vehicles.* Tech. rep., RAND Corporation. https://doi.org/10.7249/RR2664

REMUS AUV. (n.d.). Woods Hole Oceanographic Institution. Retrieved May 2025, from https://www.whoi.edu/what-we-do/explore/underwater-vehicles/auvs/remus/

REMUS UUVs. (n.d.). HHI. Retrieved June 2025, from https://hii.com/what-we-do/capabilities/unmanned-systems/remus-uuvs/

Roberts, G. N., & Sutton, R. (Eds.). (2013). *Further Advances in Unmanned Marine Vehicles.* The Institution of Engineering and Technology. https://doi.org/10.1049/PBCE002E

Rutherford, K. T. (2008). *Autonomous Underwater Vehicle Design Considering Energy Source Selection and Hydrodynamics.* University of Southampton.

Sentry. (n.d.). Woods Hole Oceanographic Institution. Retrieved June 2025, from https://www.whoi.edu/what-we-do/explore/underwater-vehicles/auvs/sentry/

Side Scan Sonar. (n.d.). NOAA Ocean Exploration. Retrieved January 2025, from https://oceanexplorer.noaa.gov/technology/sonar/side-scan.html

Stojanovic, M. (2007). Underwater Acoustic Communication: Design Considerations on the Physical Layer. *IEEE Communications Magazine.*

Sub-Bottom Profiler. (2023, June 26). NOAA Ocean Exploration. https://oceanexplorer.noaa.gov/technology/sub-bottom-profiler/sub-bottom-profiler.html

Synthetic Aperture Sonar. (n.d.). NOAA Ocean Exploration. Retrieved January 2025, from https://oceanexplorer.noaa.gov/technology/sonar/sas.html

Tiano, A., & Ferri, G. (2018). Underwater Vehicle Navigation Using SINS, DVL and LBL with Intelligent Sensor Data Fusion. *Journal of Intelligent & Robotic Systems, 91*, 263–278.

Tijjani, A. S., & Chemori, A. (2021). *Underwater Vehicles: Design and Applications.* Nova Science Publishers, Inc.

Wadoo, S., & Kachroo, P. (2011). *Autonomous Underwater Vehicles: Modeling, Control, Design, and Simulation.* CRC Press Taylor & Francis Group.

Zhang, T., Liu, B., & Liu, Y. (2018). Positioning Systems for Jiaolong Deep-Sea Manned Submersible: Sea Trial and Application. *IEEE Access.* https://doi.org/10.1109/ACCESS.2018.2881390

Zhou, C. (2021). *Design of Autonomous Underwater Vehicles's Antenna System.* Stockholm: KTH Royal Institute of Technology. https://www.diva-portal.org/smash/get/diva2:1639312/FULLTEXT01.pdf

6 Underwater Communication

6.1 REQUIREMENTS

Underwater communication technologies are critical to a wide range of scientific, industrial, and naval (military) applications. Over decades, engineers and researchers have developed specialised techniques to transmit, receive, and process acoustic data underwater.

The ocean's vast, dynamic, and harsh environment poses unique challenges for communication systems. Unlike terrestrial or satellite communication, underwater channels are characterised by high signal attenuation, variable sound speed, and complex multipath propagation. Most of the research and development in ocean acoustics and acoustic communication was driven by military applications of sonar and improvements to detect submarines and ships (Moore, Bohm, & Jensen, 2010).

The rise of autonomous underwater vehicles (AUVs) and unmanned underwater vehicles (UUVs) has further driven innovation in underwater communication (Aoki, 2008). Autonomous systems require robust, low-latency, and energy-efficient communication links to share sensor data, receive mission updates, and coordinate with other platforms (Kilfoyle & Baggeroer, 2000).

Current underwater communication technologies can broadly be categorised into three primary modes: acoustic, optical, and electromagnetic (EM). Each has unique advantages and limitations, which dictate their suitability for different applications (Kilfoyle & Baggeroer, 2000).

6.2 ACOUSTIC COMMUNICATION

Acoustic communication relies on the transmission of sound waves, which can be modulated to carry digital data. Acoustic means are the most widely used method to communicate underwater due to the relatively low absorption of sound in water. Sound travels in water at approximately 1,500 m/s, though this speed varies with temperature, salinity, and depth. Sound waves can propagate long distances, albeit with limitations imposed by multipath effects, ambient noise, and Doppler shifts (Goetz et al., 2012).

6.2.1 Acoustic Modem

The acoustic modem is the equipment used for underwater wireless communication between a UUV and its mother ship, or between a UUV and other UUVs/nodes. It

DOI: 10.1201/9781003737476-6

TABLE 6.1
Typical Performance Characteristics for Acoustic Modems

Make/Model	Power (Tx/Rx/Idle)	Data Rate	Range
Sonardyne Modem 6	–	≤9 kbps	≤5 km
WHOI Micromodem-2	Idle ~0.5 W	80–5,400 bps	~1 km+
EvoLogics S2C R	2.5 mW–65 W (Tx varies)	13.9–31.2 kbps	1–3.5 km
Water Linked M16	Rx < 0.3 W, Tx < 1.5 W	– (packets)	~1 km
LinkQuest UWM	Tx 2–8 W, Rx ~?	up to 38.4 kbps	~1–5 km
Benthos DAT	–	~0.14–15 kbps	2–6 km (extensible)

works by encoding and sending digital data (e.g., telemetry, commands, or sensor readings) as acoustic signals. A Receiver decodes the signals into usable information. It is used for communicating with submerged vehicles, for remote sensor data retrieval, and for command and control in subsea missions. There is a fundamental trade-off: lower frequency corresponds to longer range, but lower data rate (e.g., 8–15 kHz, ~1–2 kbps). Higher frequency provides shorter range, with higher data rate (e.g., 35–50 kHz, ~20–50 kbps).

Performance is affected by turbulence, noise, thermoclines, and multiple acoustic paths. Directional transducers can improve range and reduce interference. The size of the equipment can vary, with an example of a relatively large one being EvoLogics S2C-R,[1] weighing 2 to 9 kg and measuring about 110 mm (diameter) × 178 mm. A summary of typical acoustic modem equipment characteristics is given in Tables 6.1 and 6.2.

6.2.2 Acoustic System Characteristics

- **Frequency Range**: Acoustic systems typically operate in frequencies ranging from a few hundred hertz to tens of kilohertz. Lower frequencies enable long-range communication (up to tens or even hundreds of kilometres) but provide lower data rates, whereas higher frequencies support higher bandwidth but are limited in range.
- **Modulation Techniques**: Common modulation techniques include Frequency Shift Keying (FSK), Phase Shift Keying (PSK), and more advanced methods such as Orthogonal Frequency Division Multiplexing (OFDM) to maximise data throughput and reliability.

6.2.3 Performance Metrics

- **Range**: Low-frequency acoustic communication can achieve ranges from 10 to over 100 km in optimal conditions. However, in practical applications, ranges are often limited to a few kilometres in noisy or shallow environments.

TABLE 6.2
Features of Acoustic Modems for AUVs (Hwangbo, Jeon, & Park, 2013)

Parameter	Typical Range/Values	Remarks
Size	20–50 cm length, 7–15 cm diameter	Compact for AUV integration; larger models exist for long-range or high-power use
Weight	1–5 kg in air	Lighter in water due to buoyancy
Power consumption	**Transmit**: 5–50 W **Receive/Idle**: 0.5–2 W	Depends on range and data rate; low-power sleep modes are often available
Operating voltage	12–36 V DC	For compatibility with various platforms
Frequency bands	8–50 kHz (most common)	Lower freq = longer range, lower bandwidth; Higher freq = higher bandwidth, shorter range
Data rate (bandwidth)	**Short range**: up to 50–100 kbps; **Long range (5–10 km)**: ~100–1,000 bps	Bandwidth heavily depends on range, environment, and modulation
Communication range	500 m–10+ km	Line-of-sight; affected by salinity, noise, multipath, etc.
Latency	0.5–2 s typical round-trip	Due to the sound speed in water (~1,500 m/s) and processing delay
Bit error rate (BER)	10^{-5} to 10^{-8} (with error correction)	Varies with SNR and channel conditions
Supported protocols	Proprietary or standards like JANUS (NATO), WHOI Micromodem protocol	Some modems allow networking with TDMA/CSMA/Token-passing
Duplex mode	Half-duplex (mostly)	Full-duplex is rare and power-hungry
Depth rating	100–6,000 m	Based on housing material (aluminium, titanium, etc.)

- **Bandwidth**: Typical acoustic communication channels provide bandwidths from a few hundred hertz to several kilohertz, which translates to data rates in the range of a few kilobits per second to tens of kilobits per second. Field tests of modern acoustic modems have demonstrated ranges of up to 10 km in favourable conditions, with effective bandwidths ranging from 1 to 10 kbits per second. Advanced modulation schemes have pushed these limits further, with some experimental systems achieving 20 kbits per second over moderate ranges.
- **Latency**: Acoustic signals suffer from significant latency due to the slow speed of sound in water. This can be on the order of hundreds of milliseconds to seconds, depending on the distance. This is acceptable for many autonomous and monitoring applications, but limits real-time control in high-speed operations.

6.3 OPTICAL COMMUNICATION

Optical communication underwater uses light (typically lasers) to transmit data. This method can offer very high data rates over short distances. Light propagates in water with significant attenuation due to absorption and scattering. The performance is highly dependent on water clarity; it is most effective in clear, deep water (Partan, 2007).

6.3.1 Characteristics

- **Wavelengths**: Blue and green wavelengths are preferred because they experience less attenuation in seawater compared to other wavelengths.
- **Modulation**: High-speed modulation techniques enable data rates that can reach megabits per second, albeit over ranges typically limited to tens of metres.

6.3.2 Performance Metrics

- **Range**: Optical communication systems typically achieve ranges of 10–100 m in clear ocean water. In turbid or coastal waters, the effective range may be reduced significantly, often to less than 10 m (Partan, 2007).
- **Bandwidth**: Optical channels can provide bandwidths from several megahertz up to several gigahertz, supporting high data rates. Optical underwater communication systems have achieved data rates of up to several megabits per second in clear water over distances of 10–50 m (Abd El-Mottaleb, 2025).
- **Latency**: Because optical signals travel at the speed of light, latency is extremely low (a few microseconds); however, overall system latency may be affected by processing and alignment issues.

6.3.3 Applications of Optical Communication

- **Scientific**: High-speed data transfer from AUVs for real-time video, imaging, and sensor data in short-range operations.
- **Industrial**: Inspection and control of underwater infrastructure where high-resolution imagery is required.
- **Naval**: Short-range secure communication between underwater vehicles and between a vehicle and a surface station.

6.4 ELECTROMAGNETIC (EM) COMMUNICATION

Electromagnetic communication underwater is challenging due to the high attenuation of radio waves in seawater, but it finds niche applications, particularly in shallow or near-surface environments. EM waves in water are quickly absorbed, with the penetration depth inversely related to frequency. Lower frequencies can penetrate further but offer very limited bandwidth.

6.4.1 Characteristics

- **Frequency Range**: Extremely low frequency (ELF) and very low frequency (VLF) signals are used for long-range communication but support very low data rates.
- **Antennas**: Specialised antenna designs are required, often with large physical dimensions, which can be impractical for many applications.

6.4.2 Performance Metrics

- **Range**: EM communication can achieve longer ranges than optical in certain conditions (e.g., ELF can propagate for hundreds of kilometres), but effective communication for mobile platforms is typically limited to short ranges (Dhanak & Xiros, 2016).
- **Bandwidth**: The available bandwidth is very low, often only a few tens or hundreds of bits per second.
- **Latency**: Latency is minimal due to the high speed of EM wave propagation, but overall system performance is constrained by low data throughput.

6.4.3 Applications of EM Communication

- **Naval**: ELF communication has been used for secure, low data rate communication with submerged vessels where other methods are impractical.
- **Industrial**: Specialised applications may include data transfer in scenarios where acoustic and optical methods are impractical.
- **Scientific**: Limited applications in scenarios requiring extremely low bandwidth but high reliability over long distances.

6.5 COMPARATIVE SUMMARY

Performance metrics such as range, bandwidth, latency, and energy consumption are key to evaluating underwater communication systems (Stojanovic, 2007). Table 6.3 summarises performance aspects and design comparisons for underwater communication technologies (Inzartsev & Pavi, 2009).

While acoustic communication remains the workhorse for long-range data transfer underwater, optical systems offer high-speed, low-latency links for short-range applications. Electromagnetic methods, though limited by severe attenuation, find niche applications in secure communication and specific industrial scenarios. We have seen that RF, optical, and acoustic links are characterised by quite different achievable ranges and bit rates. These are summarised in Figure 6.1, which emphasises the primacy of acoustic communication underwater.

TABLE 6.3
Comparison of Performance of Communication Modes

Modality	Range	Bandwidth	Latency	Advantages	Limitations
Acoustic	1 km to >10 km (optimal)	1–20 kbits/s (typical)	100–1,000 ms	Long range; robust in various conditions	Limited bandwidth; high latency due to sound speed
Optical	10–50 m in clear water; <10 m in turbid water	Up to several Mbps	<1 ms (signal speed)	High data rates; low latency	Short range; highly dependent on water clarity
Electromagnetic (ELF/VLF)	Tens to hundreds of kilometres (ELF); limited by antenna size for mobile platforms	Tens to hundreds of bits/s	Minimal latency	Long range in very specific applications	Extremely low bandwidth; large antenna requirements

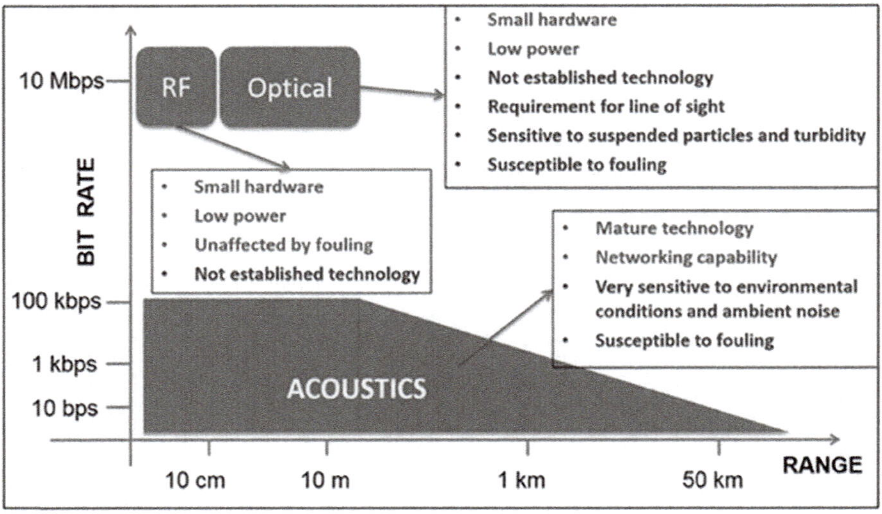

FIGURE 6.1 Achievable range and data bandwidth of different communication means in the underwater domain.

Notes: **Ferri, Munafo, Tesei, and Braca (2017); https://www.researchgate.net/publication/318213822_ Cooperative_Robotic_Networks_for_Underwater_Surveillance_an_Overview**

6.6 IMPROVING ACOUSTIC COMMUNICATION

It is not enough to just buy an acoustic modem and expect it to work uninterruptedly on a UUV. The ocean is very unlike the air (or fibre networks!), and acoustic performance is definitely not ‘plug and play’. A holistic and practical approach towards reliable acoustic communication involves several aspects: technology choices, integration hygiene, and deployment planning.

6.6.1 Technology Choices

The choice of the frequency band is a trade-off between absorption, ambient noise, and usable bandwidth. Sound attenuation in warm shallow waters increases steeply with frequency, which drives the trade-off between range and data rate when selecting an operating band.[2]

Other choices with regard to technology are transmit source levels, directivity, modulation, and coding. Transmit source levels are limited by cavitation thresholds when operating shallow or with small vehicles, and by the capability of the UUV’s power system to actually supply the transient peaks that the transducer demands.

Directivity is equally important, and directional transducers would be more energy-efficient than omnidirectional ones if the geometry of the mission can be predetermined. On the receiver side, spatial diversity can be provided by having a small array of hydrophones on the surface platform (support ship) or on a UUV that is large enough to support such a distributed array.

When selecting modulation, the highest bit rates may be less important than the robustness of communication. It is advisable to choose the lowest bit rate that still meets the mission data requirements and use forward error correction or erasure coding to overcome challenging conditions.

Navigation is also part of the early design stage, and it needs to be decided whether to have simple two-way ranging or if the system being deployed can support synchronised clocks for one-way measurements. If the set-up favours long baseline (LBL), inverted LBL, Ultra-Short Baseline (USBL), or inverted USBL navigation, that should form part of the system design for communication as well. These acoustic navigation techniques are discussed in Chapter 7, and their equipment and sensors have been mentioned in Chapter 5.

Beyond frequency, power, and modulation, there are also important choices in how multiple users share the channel and how higher-level acoustic data is handled. Based on the operational scenario, choices can be made between Medium access control (MAC), time-division scheme (TDMA), or random-access approaches (CSMA or MACA). Some data is time-sensitive (like vehicle position), while other data may be large but delay-tolerant. An acoustic modem or stack that allows the user to configure quality of service, set time-to-live values, and even declare content types (e.g., image, telemetry, audio) can make smarter transmission decisions and improve the reliability of communication.[3]

6.6.2 Integration Hygiene

Mechanical integration of acoustic sensors must consider factors such as the placement of the transducer on the UUV. Thrusters generate noise and, more critically, microbubbles that persist in the water and scatter sound, destroying coherent links. Flow noise from fast-moving vehicles, vibrations from mechanical structures, and even the rigidity of the mounting bracket all contribute to degraded performance (Ignatius & Chitre, 2026).[4]

Electrical integration is equally important to ensure reliable acoustic performance. Modems draw sharp bursts of current during transmission, and if the power bus is noisy or poorly filtered, that noise couples back into the receiver front end. Clean power supplies and proper filtering are essential to ensure consistent source levels and avoid self-interference. Once integrated, it is important to measure actual source levels in situ because cabling, mounts, or limited current draw from a vehicle's battery system may reduce output below specification.[5]

The third pillar of hygiene is software integration. Internet protocols should not be used blindly, because the assumptions of fast feedback and low loss don't hold underwater; purpose-built protocols are far better. By tagging data, applications (such as UnetStack[6]) can attach metadata such as robustness level, reliability, time-to-live, priority, and MIME type to each message, allowing the modem to make smarter delivery decisions. This gives the modem the context it needs to adapt to current conditions, compressing an image if bandwidth is poor, discarding stale telemetry instead of wasting airtime, or prioritising urgent commands.

The physical characteristics of bandwidth, absorption, and Doppler effects need to be understood and accounted for by a practitioner to effectively utilise even a proven product for specific user mission requirements.

6.7 ACOUSTIC COMMUNICATION STRATEGIES

Research into low-power hardware and communication techniques will continue to be critical. Reliable connectivity can also be improved by techniques such as adaptive communication links, which are discussed further.

6.7.1 Low-Power Communication and Computation

Advances in low-power processors, energy-efficient sensors, and adaptive modulation schemes for acoustic communications can help extend mission durations. Novel battery technologies, such as pressure-tolerant lithium-polymer batteries, are also under investigation to provide higher energy densities while maintaining reliability under extreme underwater conditions. The future of energy management lies in intelligent duty cycling—where sensors and communication modules are activated only when necessary—and adaptive sampling strategies that adjust the data collection frequency based on real-time environmental assessments. Further research will likely focus on predictive models that can forecast energy consumption patterns and adjust operational parameters dynamically, ensuring that the AUV remains within safe energy limits while still providing timely data to operators.

6.7.2 Hybrid Communication Networks

The integration of hybrid communication networks that combine acoustic, radio frequency (RF), and satellite communications will play a crucial role in future AUV operations (Rice & Green, 2008). Such networks can provide high-bandwidth links when the AUV surfaces or operates near a relay platform, while maintaining energy-efficient acoustic communications during submerged operations (Chaudhary et al., 2023). Research into protocols that seamlessly switch between these communication modes will further improve the trade-off between energy conservation and rapid data reporting.

6.7.3 Adaptive Links

Adaptive modulation schemes allow the system to step down to more robust modes when multipath or Doppler is high, and step up again when the channel clears to deliver higher throughput. Similarly, error control coding can be tuned dynamically: adding more redundancy when losses rise, and trimming it back to recover data rate when the link is clean. Erasure codes are particularly useful to transmit a little extra upfront, so the receiver can reconstruct the message even if several frames are lost, without the delay of waiting for acknowledgements.[7]

6.7.4 Adaptive Command and Control

The quality and reliability of acoustic communication with UUVs can be further enhanced by the UUV itself adapting its mission in real time based on the feedback from its communication system. For example, if the modem detects that the channel is degrading, the vehicle might pause bulk transfers, climb or dive a few metres to escape a dead zone, or slightly adjust its course to maintain better connectivity.[8]

As an autonomous vehicle moves, it can measure acoustic signal strength along its path and use those observations to predict what the field will look like in the region just ahead. Rather than relying purely on machine learning, which struggles to extrapolate beyond the data it has seen, this method embeds the underlying physics of acoustic propagation directly into the learning process (Chitre & Kexin, 2022).

6.8 CONCLUSION

Underwater communication requirements span the scientific, industrial, and naval domains. Future developments are expected to focus on hybrid communication systems that leverage AI for dynamic channel adaptation, the integration of advanced materials for improved transducer performance, and robust, energy-efficient designs that extend operational endurance. Overcoming challenges such as environmental attenuation, limited bandwidth, and latency will require continued research and innovation, supported by interdisciplinary collaboration among academia, industry, and defence organisations.

NOTES

1. https://www.evologics.com/product/s2c-r-18-34-1
2. https://blog.unetstack.net/acoustic-communication-and-navigation-for-subsea-robotics-part1
3. https://blog.unetstack.net/acoustic-communication-and-navigation-for-subsea-robotics-part1
4. https://blog.unetstack.net/acoustic-communication-and-navigation-for-subsea-robotics-part1
5. https://blog.unetstack.net/acoustic-communication-and-navigation-for-subsea-robotics-part1
6. https://subnero.com/feature/research/unetstack/2026/01/08/Building-Connectivity-with-Confidence-for-Subsea-Robotics.html
7. https://blog.unetstack.net/acoustic-communication-and-navigation-for-subsea-robotics-part2
8. https://blog.unetstack.net/acoustic-communication-and-navigation-for-subsea-robotics-part2

REFERENCES

Abd El-Mottaleb S. A., Singh, M., Atieh, A., & Aly, M. H. (2025). High Data Rate Underwater Optical Wireless Communication Systems with ICSM Codes Within Green Spectrum. *Optical and Quantum Electronics*, *57*, 213. https://doi.org/10.1007/s11082-025-08065-8

Allen, B., Stokey, R., Austin, T., Forrester, N., Goldsborough, R., Purcell, M., & Von Alt, C. (2002). REMUS: A Small, Low Cost AUV; System Description, Field Trials and Performance Results. In *Oceans '97. MTS/IEEE Conference Proceedings* (Vol. 2, pp. 994–1000). IEEE. https://doi.org/10.1109/oceans.1997.624126

Aoki, T. (2008). Urashima. In T. Ura (Ed.), *Autonomous Underwater Vehicles, Design and Operation* (2nd ed.). Ura.

Chaudhary, M., Goyal, N., Benslimane, A., Awasthi, L., Alwadain, A., & Singh, A. (2023). Underwater Wireless Sensor Networks: Enabling Technologies for Node Deployment and Data Collection Challenges. *IEEE Internet Things Journal*, *10*, 3500–3524. http://doi.org/10.1109/JIOT.2022.3218766

Chitre, M., & Kexin, L. (2022). *Physics-informed Data-driven Communication Performance Prediction for Underwater Vehicles*. 6th Underwater Communications Networking (UComms 2022). Lerici. https://arl.nus.edu.sg/wp-content/uploads/2022/09/Chitre_Physics-informed_UComms-2022.pdf

Dhanak, M., & Xiros, N. (2016). *Springer Handbook of Ocean Engineering*. Springer.

Ferri, G., Munafo, A., Tesei, A., & Braca, P. (2017). Cooperative Robotic Networks for Underwater Surveillance: An Overview. *IET Radar, Sonar and Navigation*. https://doi.org/10.1049/iet-rsn.2017.0074

Goetz, M., Nissen, I., Asterjadhi, A., Casari, P., Zorzi, M., Husøy, T., . . . van Walree, P. (2012). *Underwater Acoustic Networking Technologies*. Springer. https://doi.org/10.1007/978-1-4471-2926-2

Hwangbo, S.-H., Jeon, J.-H., & Park, S.-J. (2013). Wireless Underwater Monitoring Systems Based on Energy Harvestings. *Sensors & Transducers, 18*, 113–119. https://www.researchgate.net/publication/290078574_Wireless_Underwater_Monitoring_Systems_Based_on_Energy_Harvestings/

Ignatius, M., & Chitre, M. (2026, January 8). Building Connectivity with Confidence for Subsea Robotics. *SubNero*. https://subnero.com/feature/research/unetstack/2026/01/08/Building-Connectivity-with-Confidence-for-Subsea-Robotics.html

Inzartsev, A., & Pavi, A. (2009, January 1). AUV Application for Inspection of Underwater Communications. In A. V. Inzartsev (Ed.), *Underwater Vehicles* (pp. 371–398). Rijeka: InTech. https://doi.org/10.5772/6704

Kilfoyle, D. B., & Baggeroer, A. B. (2000). The State of the Art in Underwater Acoustic Communication. *IEEE Journal of Oceanic Engineering*, *25*(1), 4–27. https://doi.org/10.1109/48.820733

Moore, S., Bohm, H., & Jensen, V. (2010). *Underwater Robotics: Science, Design and Fabrication*. Monterey, CA: Marine Advanced Technology Education Center.

Partan, J. K. (2007). A Survey of Practical Issues in Underwater Networks. *ACM SIGMOBILE Mobile Computing and Communications Review*, *11*(4), 23–33. https://doi.org/10.1145/1347364.1347372

Randeni, S., Schneider, T., Bhatt, E. C., Víquez, O. A., & Schmidt, H. (2022, November). A High-resolution AUV Navigation Framework with Integrated Communication and Tracking for Under-ice Deployments. *Journal of Field Robotics*. https://doi.org/10.1002/rob.22133

Rice, J. A., & Green, D. (2008). *Underwater Acoustic Communications and Networks for the US Navy's Seaweb Program*. Sensor Technologies and Applications SENSORCOMM '08. San Diego. https://doi.org/10.1109/SENSORCOMM.2008.137

Stojanovic, M. (2007). Underwater Acoustic Communication: Design Considerations on the Physical Layer. *IEEE Communications Magazine*.

7 Guidance, Navigation, and Control

Navigating beneath the surface of the water poses a unique set of challenges. This chapter provides an overview of underwater navigation methods used in modern AUVs. The concepts of guidance, navigation, and control (GNC) are described. The mathematical model of motion is linked to control theory. GNC translates mission-level commands into actuator-level signals. High-level guidance algorithms plan missions and generate trajectories for the low-level controllers. The benefits of sensor fusion and processing improvements for reducing drift and increasing navigation accuracy are discussed, with accuracy figures from various field trials.

7.1 FUNDAMENTALS

Navigating beneath the surface of the water poses a unique set of challenges. Since radio-frequency signals attenuate rapidly in water, underwater systems cannot rely on the Global Positioning System (GPS) or on radio communication like land or aerial vehicles. The many challenges underwater include sensor drift, multipath effects in acoustics, and the highly dynamic nature of the underwater environment.

Instead, modern AUVs need to use a combination of inertial sensors, acoustic ranging systems, Doppler velocity logs (DVLs), and occasionally geophysical cues to estimate their position.

For clarity, the components of the Guidance, Navigation, and Control (GNC) subsystem (De Barros, Pascoal, & De Sa, 2004) of a UUV comprise the following components:

- **Guidance**: Determines where the vehicle should go, including path planning, waypoint navigation.
- **Control**: Determines how the vehicle moves, i.e., depth control, heading stabilisation, thruster commands.
- **Navigation**: Estimates position from sensors such as IMU, DVL, sonar.

7.1.1 Control Techniques

Numerous control methods have been proposed for AUVs to achieve high performance despite nonlinear dynamics, uncertainties, and environmental changes. The control system should be able to learn and adapt to changes in vehicle dynamics and the environment. Various control strategies and algorithms are listed below.

 DOI: 10.1201/9781003737476-7

7.1.1.1 Classical/Model-Based Control

- **PID Control**: Utilises sensor feedback and processes it proportionally, integrally, and derivatively. PID control has been successfully implemented, but is better suited to linear models and may not provide the same performance as nonlinear schemes for complex systems.
- **Computed Torque Control**: A control law based on a mathematical model of the system. Simple in design, but stability is sensitive to parameter variations and sensor noise.
- **Nonlinear Control**: Essential due to the highly nonlinear dynamics of underwater vehicles. Techniques like Backstepping Integral Adaptive (BIA) control stabilise the vehicle in position and orientation.
- **H-infinity Robust Control**: Aims to achieve robust control performance, ensuring internal stability and disturbance attenuation. Can be used for path-following control.

7.1.1.2 Adaptive and Robust Control

These methods focus on handling model uncertainty and disturbances.

- **Adaptive Strategies**: Account for uncertainty in the model by adapting control parameters. An adaptive control law based on Euler angle representation has been proposed. Adaptive (integral) control laws help compensate for hydrodynamic effects. Adaptive computed torque control can withstand higher disturbances and compensates parameter variations.
- **Robust Approach**: Designed to be insensitive to parameter variations and disturbances. Sliding mode control (SMC) is described as a robust approach.
- **Sliding Mode Control (SMC)**: Uses a mathematical technique called sliding mode to achieve robust control regardless of dynamic variations or environmental disturbances. Robust to parameter uncertainties. SMC, employing concepts of sliding surfaces and boundary layers, achieves an optimal trade-off between control bandwidth and tracking precision. Coupled SMC has an advantage over uncoupled SMC for trajectory tracking when multiple DOFs are excited.

7.1.1.3 Intelligent and Learning-Based Control

- **Neuro-fuzzy Intelligent Control**: Seen as an option to provide real-time online modelling or compensation for poorly understood vehicle dynamics or unknown parameters. Examples include self-adaptive neuro-fuzzy inference systems.
- **Machine Learning (ML)**: Used for tasks like adjusting AUV trajectory, herd localisation, and can develop modified strategies for behaviour selection. Robot learning techniques can improve autonomy and robustness. Gaussian process (GP) regression, a machine learning method, can be used to create surrogate models for predicting complex hydrodynamic interactions in real time.

7.2 MATHEMATICAL MODEL OF MOTION

The mathematical model of the motion of an Autonomous Underwater Vehicle (AUV) correlates its rigid-body dynamics, hydrodynamics, mass distribution, and external forces. This model is typically a set of nonlinear, coupled differential equations (often 6-DOF). This model forms the foundational basis for the design of the AUV's Guidance, Navigation, and Control (GNC) systems. This model describes how the vehicle responds to control inputs and environmental forces, using physical laws (Newton/Euler equations) adapted for underwater dynamics, including effects like buoyancy, drag, added mass, and Coriolis forces.

This is formally expressed through six coupled nonlinear differential equations—three for linear motion (surge, sway, heave) and three for angular motion (roll, pitch, yaw).

Solving these coupled equations over discrete time steps allows prediction of vehicle trajectories under varying conditions of fin deflection, thrust, and disturbance forces. Numerical integration methods (such as fourth-order Runge–Kutta) are commonly used to solve these equations in simulation. This enables trajectory prediction in the three-dimensional underwater domain, accounting for the interactions between control inputs, hydrodynamic reactions, and inertial responses.

7.2.1 Hydrodynamic Coefficients

In the mathematical modelling of an Autonomous Underwater Vehicle (AUV), hydrodynamic coefficients are essential parameters that quantify how fluid dynamic forces and moments act on the vehicle as it moves through water. These coefficients encapsulate the effects of added mass, damping, lift, drag, and restoring forces, linking the vehicle's velocity, acceleration, and control inputs to its dynamic response. They are indispensable in both the design and control phases, allowing engineers to simulate motion accurately, predict performance, and develop robust control algorithms.

Hydrodynamic coefficients can be broadly categorised into several groups: added-mass coefficients, which represent the inertia of water accelerated along with the vehicle; damping coefficients, which account for resistance forces opposing motion; and cross-coupling coefficients, capturing interaction between different motion degrees of freedom (for example, how a yaw motion induces sway forces).

Traditionally, these coefficients have been estimated through experimental methods such as planar motion mechanism (PMM) tests or free-running tank trials. However, with advances in computational fluid dynamics (CFD) and system identification algorithms, it has become increasingly feasible to derive them from virtual simulations combined with sensor data from prototype vehicles (Hong, Wang, & Zhang, 2024). For instance, methods using extended and unscented Kalman filters have proven effective in estimating coefficients in real time, while CFD-based models can now predict all required hydrodynamic coefficients from a single virtual free-running test with high accuracy (Ray et al., 2016).

Each hydrodynamic coefficient represents a specific physical effect, such as lift per unit yaw rate or damping per surge velocity. Understanding their significance and sensitivity aids in hydrodynamic optimisation, and in designing hull shapes or appendages for manoeuvrability and control.

Hydrodynamic coefficients, therefore, serve as the backbone of the AUV's dynamic model, translating flow physics into practical parameters that guide simulation, prediction, and control. The significance of hydrodynamic coefficients lies in their ability to bridge fluid mechanics and control theory. Their accurate estimation ensures that the mathematical model reliably mirrors real-world behaviour, which is vital for performance prediction, control tuning, and robust mission execution in complex underwater environments.

7.2.2 Linking with Control Theory

The hydrodynamic mathematical model is the unifying element enabling 'model-based' GNC design, simulation, and testing. The accuracy and completeness of the mathematical model directly determine the performance limits and safety of guidance, navigation, and control algorithms for AUVs. They transform complex hydrodynamic interactions into manageable mathematical representations, enabling model-based control strategies such as sliding-mode, LQR, or adaptive controllers to stabilise and navigate AUVs autonomously.

Advanced control strategies, including linear quadratic Gaussian (LQG) and sliding-mode controllers, rely on accurate hydrodynamic models. They compensate for nonlinearities and uncertainties by adjusting fin deflections dynamically to achieve desired motion profiles. The success of such systems depends fundamentally on the fidelity of hydrodynamic data linking hull configuration and control surface action to the resulting dynamic behaviour.

Model-based GNC design cycles through model simulation, controller design, and experimental validation—each stage refining both the mathematical model (to better match real dynamics) and the GNC architecture for precision, robustness, and safety. Thus, the mathematical model is the foundation that structurally links AUV physics to every facet of motion planning, perception, and actuation.

7.2.2.1 Guidance

The mathematical model predicts how different sequences of control inputs will change the AUV's trajectory and orientation; path planners and trajectory generators use the model to ensure feasible and safe motion.

Guidance algorithms require predictions of vehicle trajectories based on possible control inputs. The mathematical model simulates the AUV's response to waypoints, paths, and constraints, enabling path planning, i.e., determining feasible and energy-efficient paths. It also enables simulation for manoeuvring, essential for predicting turn radii, minimum depths, and collision avoidance.

7.2.2.2 Navigation

State estimators like Kalman filters (KF) combine model-predicted states with real-time sensor measurements to estimate the true position, velocity, and attitude of the AUV, accounting for uncertainty and disturbances.

Navigation filters like KF, extended KF, and particle filters blend model predictions with sensor data (INS, DVL, GPS, acoustics) to estimate the true state (position, velocity, attitude) of the AUV. The model provides the predicted evolution of state (process equations), which are essential for robust state estimation in the underwater environment where sensor access is intermittent.

7.2.2.3 Control

Control system design can use classical PID, LQR, sliding mode, or modern adaptive/model predictive control. All of these require accurate models for the following:

- Linearise the system around operating points (for controller synthesis).
- Simulate closed-loop performance (response to commands, disturbances).
- Tune controllers (gain, stability margins).
- Guarantee safe and stable operation by accounting for nonlinearities and coupling effects.

Control laws such as PID or model-based controllers are derived using simplified linearised or full nonlinear models of the vehicle, ensuring stability and desired tracking behaviour under anticipated environmental and actuation conditions.

7.3 CONTROL STRATEGIES

In a layered AUV control architecture, guidance and control algorithms typically operate at the control layer, which sits between high-level mission logic and low-level device drivers. The GNC translates *mission-level commands* into *actuator-level signals.*

Control strategies refer to algorithms or techniques that drive decision-making and actuation in autonomous systems. These strategies transform intent (goals) into action (movement or behaviour) while dealing with constraints, uncertainty, and changing environments.

Control strategies are broadly classified into high and low-level strategies.

- High-Level Control Strategies:
  - Focus on mission planning, decision-making, behaviour arbitration, and task sequencing.
  - Typically operate at the task layer.
  - Tend to be deliberative or reactive logic governing what to do next.
- Low-Level Control Strategies:
  - Focus on motion control, stabilisation, tracking, and actuator coordination (Fossen, 1991).
  - Typically operate at the control layer.
  - Tend to be mathematical or feedback-based algorithms for how to move or behave.

AUV control systems combine both low-level and high-level control strategies. ROS (1 and 2) and MOOS IvP both support high-level control strategies. JAUS provides a standardised interface for control commands, but only supports messaging for high/low-level control. Data distribution service (DDS) does not natively support high/low-level control, but supports comms for any custom control architecture.

The subsequent sections discuss high-level control strategies, followed by low-level control strategies for AUVs.

7.3.1 High-Level Guidance and Decision-Making

High-level guidance algorithms plan missions and generate trajectories for the low-level controllers. They are located at the *'Task'* layer. They decide *what* to do, *when* to do it, and *how* to recover from faults. They operate at timescales of seconds to minutes. Examples are as follows:

- **Waypoint Navigation**: Involves planning a path that connects predefined waypoints. This is common in survey missions.
- **Behaviour-Based Control**: Decomposes complex missions into simpler behaviours (e.g., 'obstacle avoidance', 'data collection', 'energy conservation') that are triggered based on environmental cues.
- **Fuzzy Logic Control**: Uses linguistic rules to handle uncertainty and non-linear dynamics in the decision-making process.
- **Artificial Intelligence (AI) and Machine Learning (ML)**: Emerging techniques that allow for autonomous adaptation and mission re-planning based on sensor data, historical performance, and environmental changes.

Salient high-level control strategies are compared in Table 7.1.

Finite State Machines (FSMs) and IvP Helm are mature and widely deployed in fielded UUVs. Behaviour Trees offer a more maintainable and scalable alternative to FSMs. AI/ML-based planning is still emerging but increasingly valuable in adaptive autonomy. Swarm coordination strategies are vital in next-gen UUV systems operating in fleets or MCM (mine countermeasure) missions.

7.3.2 Low-Level Control Strategies

These algorithms operate in the inner control loops, maintaining vehicle stability and following short-term trajectories. They operate at timescales of milliseconds to seconds. They are located at the control layer. Their functions include stabilising depth, heading, and speed (Fossen, 1991). Examples of such algorithms are mentioned below:

- **PID Controllers**: Widely used due to their simplicity and effectiveness. They provide corrective feedback for errors in depth, heading, and speed.

TABLE 7.1
Summary of High-Level Control Strategies for AUVs

Control Approach	Type	Key Features	Typical Applications (for UUVs)
Finite state machines (FSM)	Rule-based/discrete logic	Simple, deterministic transitions between states	Mission sequencing (e.g., descend → survey → surface)
Behaviour trees	Modular reactive planning	Hierarchical, modular, readable, interruptible	Flexible mission planning, adaptable behaviours
IvP helm (MOOS-IvP)	Multi-objective behaviour arbitration	Optimises weighted behaviours at runtime, supports dynamic priorities	Line following + collision avoidance + search; emergency recovery
Hierarchical task networks (HTN)	Symbolic planning	Decomposes tasks into subtasks recursively, logic-driven	Complex mission plans, automated planning in structured scenarios
Deliberative planning	Search-based planning (e.g., A, D, RRT)	Graph-based planning, maps the environment, generates optimal paths	Pre-mission path planning, obstacle-aware navigation
Reactive planning (e.g., VFH, DWA)	Real-time reactive control	Uses sensor data for local obstacle avoidance in real time	Terrain following, dynamic obstacle avoidance
AI/ML-based planning	Learning-based autonomy	Learns mission policies or decision models from data or simulations; allows for autonomous re-planning	Adaptive re-tasking, mission optimisation in unknown conditions
Swarm/fleet coordination	Distributed task allocation	Assigns tasks across multiple vehicles, often using auctions or consensus	Multi-UUV mapping, coverage, cooperative search
Goal-oriented action planning (GOAP)	Goal-driven symbolic planning	Flexible plans generated from goals and the current world state	Dynamic goal execution and re-planning under environmental changes
Waypoint navigation		Pre-planned mission paths; simple geometric path planning	Survey missions, environmental monitoring
Fuzzy logic control	Rule-based/heuristic	Handles vague or imprecise models, intuitive tuning	Buoyancy control, variable ballast systems; Nonlinear control scenarios, decision-making under uncertainty

- **Sliding Mode Control (SMC)**: Offers robustness in the presence of model uncertainties and external disturbances, making it well suited for harsh underwater environments.
- **Model Predictive Control (MPC)**: Uses an explicit model of the vehicle dynamics to optimise control actions over a prediction horizon, thereby handling multivariable constraints and nonlinearities.
- **Adaptive Control**: Adjusts controller parameters in real time to cope with changing dynamics, such as variations in drag or payload.
- **Behaviour-Based Control** often acts as a bridge between high-level decisions and low-level actuation, especially in reactive systems.
- **AI/ML-Based Methods** are becoming popular for adaptive or complex environments, though their trustworthiness and certifiability are active research areas in safety-critical UUV applications.

Table 7.2 summarises common control strategies used in AUVs, along with key features and typical applications.

7.4 UNDERWATER NAVIGATION TECHNIQUES

Continuous improvements in sensor technologies and sophisticated sensor fusion algorithms have significantly enhanced the accuracy and robustness of underwater navigation. However, it still remains a challenge compared to the accuracy and ease of navigation on land or in the air.

A summary of localisation and navigation methods, sensors and approaches is shown in Figure 7.1. (González-García et al., 2020).

Recent field trials and experimental campaigns have demonstrated error drifts as low as 0.1–0.2% of the distance travelled. Accuracy of even 0.02% of distance travelled has been reported in a multi-week mission over a range of 1,200 nautical miles (Kongsberg Discovery, 2024). A fusion of an inertial navigation system (INS) with a DVL and acoustic updates has demonstrated drift corrections that yield final position errors in the low tens of metres over missions spanning several kilometres (Paull et al., 2014). Processing techniques such as the Extended Kalman Filters (EKF), Unscented Kalman Filters (UKF), and graph-based SLAM (Simultaneous Localisation and Mapping) have played a crucial role in minimising cumulative errors in navigation.

The subsequent sections provide an overview of the major techniques and instruments employed in underwater navigation.

7.4.1 Inertial Navigation and Dead-Reckoning

Inertial Navigation Systems (INS) are expected to provide continuous navigation information without external references. However, due to sensor noise, bias, and the integration process, errors (or drift) accumulate over time. In typical AUV missions without external correction, the INS drift may reach tens or even hundreds of metres over several hours. When operating solely on an INS (i.e., dead-reckoning), errors

TABLE 7.2
Low-Level Control Strategies for AUVs

Control Approach	Type	Key Features	Typical Applications (for UUVs)
PID control	Feedback (linear)	Simple, widely used, good performance in stable conditions; tuning required	Depth, heading, speed control, basic trajectory following
Backstepping control	Nonlinear control	Lyapunov-stable, handles system nonlinearities	Fin-actuated control, roll/pitch stabilisation
Sliding mode control	Robust control	Strong robustness to disturbances, discontinuous control	Harsh environments, actuator fault tolerance
Linear quadratic regulator (LQR)	Optimal control	Optimal trade-off between control effort and error	Smooth path tracking under linear approximations
Adaptive control	Online parameter estimation	Adjusts to changing dynamics or uncertainties	Payload changes, long missions with evolving drag/mass
Model predictive control (MPC)	Predictive optimal control	Future-aware, constraint handling, high precision; computationally intensive	Obstacle avoidance, advanced manoeuvring in cluttered environments; energy management
$H\infty$ control	Robust optimal control	Worst-case optimisation under uncertainty	Military or critical missions with unknown disturbances
Feedback linearisation	Nonlinear control	Cancels out nonlinear terms to linearise dynamics	High-performance control with accurate models
Behaviour-based control	Hybrid/reactive architecture	Uses modular behaviours (e.g., avoid, follow, search), often with arbitration	Terrain following, object tracking, reactive obstacle avoidance
AI/ML-based control	Data-driven/ learning-based	Learns control policies from data or simulation (e.g., RL, imitation learning)	Adaptive navigation, unstructured environments, mission re-planning

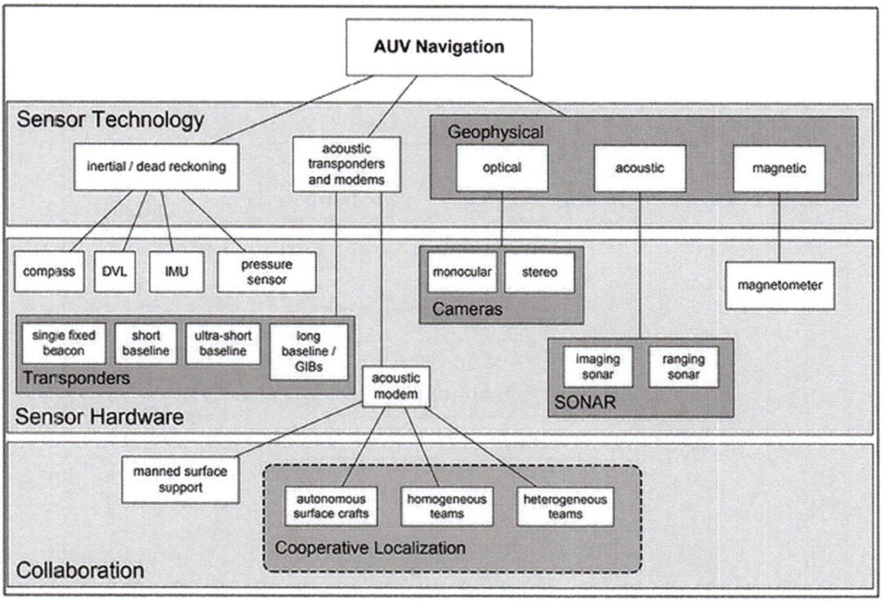

FIGURE 7.1 Localisation and navigation technologies for AUVs.

Source: **Based on Paull et al. (2014).**

accumulate with time. Without external correction, field trials have reported drifts on the order of several per cent of the distance travelled. For example, a long-duration mission spanning 10 km may accumulate drift errors on the order of 100–200 m if relying on INS alone (Kinsey, Eustice, & Whitcomb, 2006). Therefore, INS-only navigation is generally acceptable only for short-duration missions (e.g., in confined laboratory tanks or short coastal surveys).

7.4.2 Kalman Filter-Based Processing

In order to fuse the data from the INS and the aiding sensors, some form of filtering must be implemented. This is typically accomplished using an error-state Kalman filter (KF). A wide range of position aiding tools can be integrated, as well as additional velocity aiding sources (Paull et al., 2014).

A simplified schematic diagram showing the integration of DVL data is shown in Figure 7.2. (Hegrenæs et al., 2016). The KF input is the difference between the output from the appropriate aiding sensors and the INS. The output from the KF includes estimates of the slowly varying systematic errors of the navigation sensors, as well as sea current when using DVL water-track data.

7.4.2.1 Extended Kalman Filters (EKF)

EKF is one of the most widely used algorithms for sensor fusion in underwater navigation. By linearising the nonlinear models of vehicle dynamics around the current

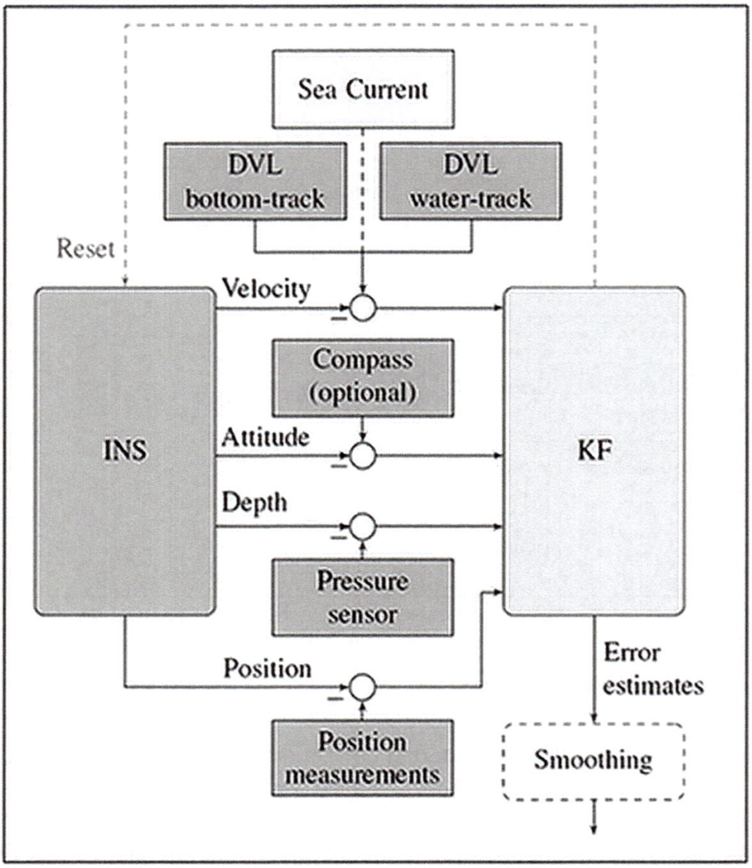

FIGURE 7.2 Simplified structure of the AUV INS with error state KF.

Notes: **Hegrenæs, Ramstad, Pedersen, and Velasco (2016); https://www.nortekgroup.com/assets/documents/Validation-of-a-New-Generation-DVL-for-Underwater-Vehicle-Navigation.pdf**

estimate, the EKF can recursively fuse INS, DVL, and acoustic measurements to produce an improved estimate of position and velocity. Field trials have shown that EKF-based INS/DVL systems can reduce drift rates to around 0.1–0.2% of the distance travelled.

7.4.2.2 Unscented Kalman Filters (UKF)

The UKF avoids linearisation by propagating a set of sample points (sigma points) through the nonlinear model. In many experiments, UKF has demonstrated a modest improvement over EKF—for example, reducing errors by an additional 10–15% in tests comparing the two methods. In one set of trials, UKF-based fusion reduced the drift error by roughly 0.1 m compared to EKF over short-duration manoeuvres.

7.4.2.3 Acoustic Positioning Systems

Acoustic positioning systems—such as long baseline (LBL), ultra-short baseline (USBL), and short baseline (SBL) systems—rely on acoustic transponders and receivers to determine the vehicle's position by measuring the time-of-flight of sound waves (Kongsberg Maritime, 2023). These systems provide absolute position fixes that can be used to correct INS drift. LBL systems use a network of transponders placed on the seafloor, whereas USBL/SBL systems use surface vessels or moored buoys. Reported accuracies vary with configuration; typical absolute positioning errors of 5–15 m are common in many field trials. These have been discussed in Chapter 5.

7.4.3 Geophysical Navigation

Geophysical navigation utilises natural environmental gradients—such as bathymetry, geomagnetic anomalies, or gravity fields—to estimate the vehicle's position by matching sensor observations to preexisting maps (Meduna, Petillot, & Lane, 2008). While geophysical methods can potentially provide long-term drift correction, their accuracy depends on the resolution and quality of the environmental map, as well as the sensitivity of the sensors used.

Using the spatial and temporal distribution characteristics of the seafloor topography and the underwater magnetic field as navigation beacons, an optimised positioning algorithm framework can be constructed. This could use EKF to couple the topographic elevation information and magnetic field strength information, with the seafloor topography and the marine geomagnetic map. This can be used to determine the relative position of the AUV with respect to the marine geophysical information chart and to correct the accumulated positioning error of the dead-reckoning (Yu et al., 2023).

When high-resolution bathymetric or magnetic maps are available, geophysical navigation methods can correct for INS drift, with reported accuracies varying from a few metres to tens of metres, depending on the feature resolution (Figure 7.3) (Teixeira & Pascoal, 2007).

7.4.4 Optical and Visual-Based Navigation

In clear water conditions, visual navigation methods using cameras or lasers can provide high-resolution, short-range measurements of the environment (Caccia, 2006). Techniques such as visual odometry and simultaneous localisation and mapping (SLAM) have been successfully applied in controlled environments (e.g., in lakes or clear ocean regions). However, in turbid or deep-water environments, optical methods are limited by light attenuation and scattering.

Despite the success of SLAM in terrestrial environments, its adaptation to underwater domains presents significant challenges. These include low visibility due to light absorption and scattering, sensor noise, dynamic lighting conditions, and distortions caused by the water column. Unlike in air, where visual and LiDAR-based

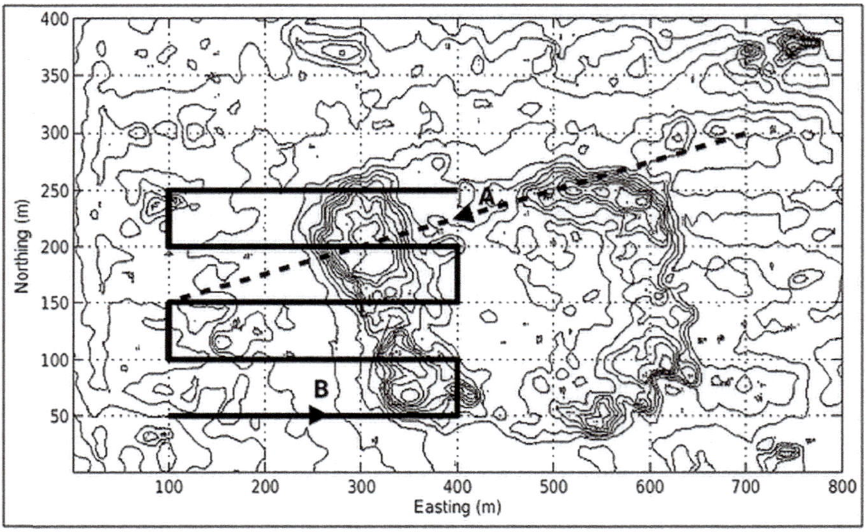

FIGURE 7.3 Localisation of trajectories based on bathymetric contour map.

Notes: **Teixeira and Pascoal (2007); https://www.sciencedirect.com/science/article/pii/S1474667015320814**

SLAM systems are widely used, underwater SLAM relies on a combination of sonar, acoustic signals, inertial sensors, and vision-based approaches, each with inherent limitations. These constraints degrade the accuracy of feature extraction, loop closure detection, and trajectory estimation, which are critical components of SLAM pipelines.

In clear water conditions, visual SLAM systems have reported relative localisation errors on the order of 20–60 cm in controlled environments; however, these figures deteriorate rapidly in turbid or dynamic settings.

7.4.4.1 Particle Filters and Graph-Based SLAM

Methods such as particle filters and graph-based SLAM provide robustness in highly nonlinear, non-Gaussian environments, enabling loop closure and long-term drift correction. Although computationally demanding, they are particularly useful in extended missions where error accumulation is a major concern.

Particle filters have been used in underwater navigation to handle non-Gaussian noise and nonlinear dynamics. Although computationally more intensive, particle filters can better manage uncertain and intermittent sensor updates (e.g., when DVL measurements are partially lost). In simulation studies, particle filter-based fusion methods have reduced cumulative errors significantly compared to traditional EKF approaches, especially in scenarios with dynamic acoustic conditions.

For applications requiring simultaneous localisation and mapping (SLAM), graph-based approaches have shown good performance in reducing drift by optimising the entire trajectory through loop closure detection (Eustice et al., 2005). In controlled field trials, graph-based SLAM has achieved relative errors on the order of 20–60 cm over short segments. When integrated with INS/DVL systems for long-term navigation, graph-based methods help correct accumulated drift by re-aligning the trajectory with previously mapped features, yielding overall position errors on the order of a few metres (Yuan et al., 2017).

7.4.4.2 Deep Learning-Based SLAM

Traditional SLAM techniques face significant obstacles, including poor visibility, dynamic lighting conditions, sensor noise, and water-induced distortions, all of which degrade the accuracy and robustness of underwater navigation systems. Recent advances in deep learning (DL) have introduced solutions to overcome these challenges.

Figure 7.4 gives an overview of sensors and methods used in underwater SLAM applications (Heshmat et al., 2025). It shows a summary of sensors, front-end VO methods, loop closure detection techniques, and back-end state estimation algorithms (Zhang et al., 2022).

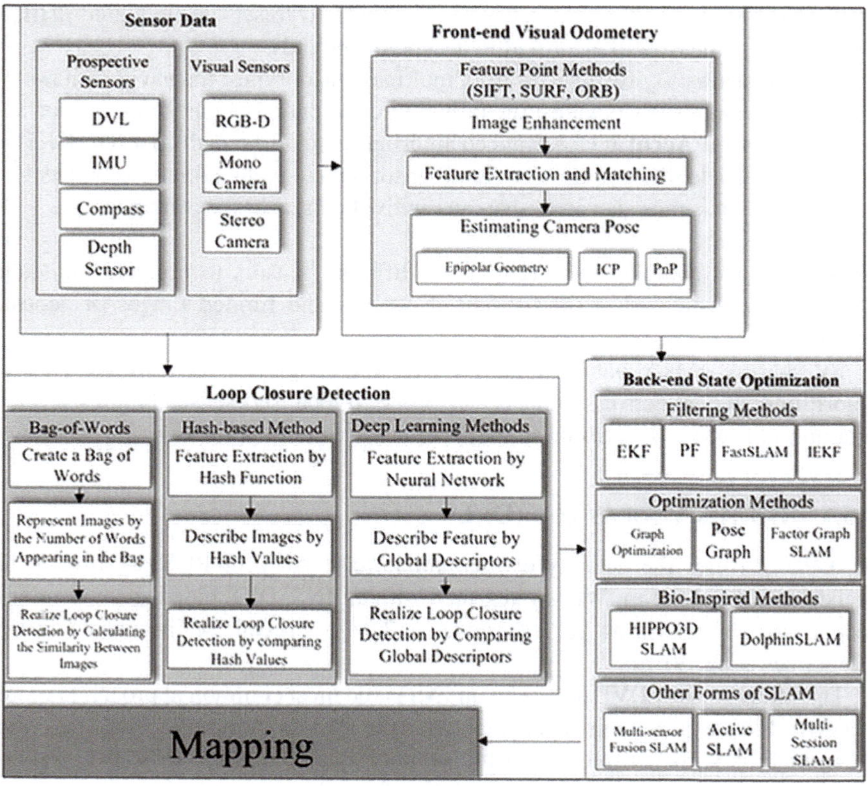

FIGURE 7.4 Overview of sensors and methods used in underwater SLAM applications. *Source*: Heshmat et al. (2025).

Notes: **https://www.mdpi.com/1424-8220/25/11/3258**

DL techniques enhance underwater SLAM by improving feature extraction, image de-noising, distortion correction, and sensor fusion. Convolutional neural networks (CNNs) and transformer-based architectures have demonstrated remarkable improvements. Multimodal approaches integrating vision, sonar, and inertial data have proven effective in compensating for sensor deficiencies, enabling more resilient localisation and mapping systems.

7.5 SENSOR FUSION AND PROCESSING

One of the main avenues for achieving high accuracy in underwater navigation is the effective integration of multiple sensors using advanced fusion techniques. The inherent drift of INS-only systems makes it imperative to integrate external sensors that provide absolute or relative fixes. Sensor fusion combines the short-term accuracy of INS with the long-term stability of external references (e.g., DVL, acoustic positioning, and geophysical measurements). The benefits include the following:

- **Drift Correction**: Periodic updates from acoustic systems or GPS (when at the surface) allow the navigation solution to 'reset' accumulated drift, keeping the overall error within acceptable bounds.
- **Robustness**: By fusing data from multiple sensor types, the navigation system is more robust against the failure or degradation of any single sensor.
- **Increased Accuracy**: Advanced filtering techniques such as EKF, UKF, and particle filters optimally blend sensor data to yield a final position estimate that is more accurate than any individual sensor measurement.

These methods combine the strengths of different sensors to mitigate individual weaknesses, particularly drift in inertial sensors and limited ranges of acoustic systems.

By sharing sensor data among vehicles and employing decentralised fusion algorithms, the collective system is less prone to localised sensor failures or drift. Examples of sensor data fusion for navigation are examined further.

7.5.1 Doppler Velocity Log (DVL)

A DVL measures the velocity of the vehicle (e.g., AUV or ROV) relative to the seabed or water column. When integrated with an INS, DVL data can substantially reduce drift by providing velocity updates. This is crucial for inertial navigation (Hegrenæs et al., 2016).

The integration of DVL data with an INS is the most common approach to reducing drift. Field experiments report that when the DVL is 'bottom-locked' (i.e., when the vehicle is sufficiently close to the seafloor), the combined INS/DVL solution can achieve drift rates as low as 0.1–0.2% of the distance travelled. For instance, a combined INS/DVL system may achieve a navigation error of around 10–20 m over a 10 km mission (Randeni et al., 2022). In one untethered under-ice mission (approximately 11 km), the majority of the navigation updates had corrections below 15 m (Bhatt, Víquez, & Schmidt, 2022).

7.5.2 INS/Acoustic Positioning Fusion

Acoustic positioning systems such as LBL, USBL, or SBL are used to provide absolute position fixes that can be fused with INS data to correct drift (Tiano & Ferri, 2018). Fusion with acoustic positioning systems is particularly beneficial in scenarios where the vehicle operates away from the seafloor (and therefore may lose DVL lock) or in dynamic environments where environmental drift (such as ice drift) must be compensated. The acoustic updates help to constrain the error growth when the vehicle is operating outside the range of the DVL (Asai et al., 2000).

Field trials in under-ice environments have been reported in the Beaufort Sea (ICEX20) over an 11 km mission. By using a combined INS/DVL/acoustic update system (HydroMAN-ICNN framework), 59% of navigation correction updates were found to be less than 15 m, and 77% of updates were below 20 m error (Randeni et al., 2022).

In other long-range missions of up to 60 km by Autosub3 beneath the Antarctic ice shelf, the incorporation of USBL updates reduced cumulative errors to within 20 m, although such performance is highly dependent on environmental conditions (Bhatt, Viquez, & Schmidt, 2022; McPhail, 2009).

7.5.3 Integration of UWSN

In addition to enhancing SLAM, an emerging research direction is the integration of underwater wireless sensor networks (UWSNs). UWSNs provide a distributed sensing framework where multiple AUVs and sensor nodes collaboratively share localisation and mapping data via acoustic communication (Chaudhary et al., 2023). By incorporating UWSN-based SLAM, navigation accuracy and robustness can be improved, especially in large-scale, multi-agent underwater operations.

7.5.4 Improvements in Processing

Advances in processing hardware and algorithms have further contributed to reducing navigation errors:

- **Real-Time Processing**: With the advent of high-performance embedded processors and Graphics Processing Units (GPUs), sensor fusion algorithms that were once too computationally demanding can now be run in real time on AUVs (Bachmayer et al., 2008).
- **Adaptive Filtering**: Modern fusion algorithms often include adaptive elements that adjust the filter's parameters in response to changes in sensor quality or environmental conditions. Adaptive filtering algorithms can adjust noise covariance matrices in real time based on sensor performance and environmental conditions. This dynamic tuning helps maintain high

accuracy even when some sensors degrade temporarily (e.g., when a DVL loses bottom-lock momentarily).

- **Outlier Rejection**: Robust statistical techniques and machine learning methods have been integrated to detect and reject erroneous sensor measurements before they corrupt the fusion process. Outlier rejection algorithms—such as those based on statistical methods or machine learning (e.g., using Voronoi diagrams for USBL data)—ensure that erroneous measurements do not corrupt the sensor fusion process. By discarding spurious data, the fusion filter can maintain a tighter bound on drift and error accumulation.
- **Multi-sensor Redundancy**: Combining measurements from multiple sensors (e.g., fusing acoustic data with visual and inertial measurements) provides redundancy that reduces the overall uncertainty and drift. Redundancy in sensor systems (e.g., having both upward- and downward-looking DVLs, or multiple acoustic transducers) further contributes to improved reliability and accuracy. In several field trials, this redundancy has allowed systems to 'reset' accumulated errors periodically, effectively capping the maximum drift over long missions.

7.5.5 Navigation Accuracy in Field Trials

In recent years, numerous field trials and experimental campaigns have been conducted to assess the navigation accuracy of modern AUVs. Although the results vary depending on the mission, environment, and sensor configuration, several common themes emerge:

Table 7.3 summarises several key field trial experiments and the navigation accuracy achieved by various AUV systems. These experiments include both single-vehicle deployments and multi-agent projects that employ sensor fusion to mitigate drift. The figures shown are representative of published field trials and engineering tests (Liu & Bucknall, 2019). Reported values are approximate and depend on environmental conditions, mission profiles, and sensor configurations (Leonard & Bahr, 2016).

European projects such as SWARMs and CoCoRo have demonstrated that in multi-agent systems, distributed sensor fusion not only allows for cooperative mission planning and obstacle avoidance, but also provides robust long-term drift correction. In these projects, the combination of INS, DVL, acoustic ranging, and environmental sensors (e.g., CTD sensors) yielded overall position accuracies typically within 15–20 m over multi-hour missions of several kilometres. Accordingly, the various systems available to an AUV for determining its position are summarised and compared in Table 7.4. (Alexandris, Papageorgas, & Piromalis, 2024).

7.6 CONCLUSION

Designing a manoeuvrable AUV hull requires harmonising hull streamlining with control effectiveness. Properly modelled hydrodynamic forces and moments, derived

TABLE 7.3
Examples of Navigation Accuracy from Field Trials

Project/Experiment	Sensor Suite	Opera-tional Range	Mission Endu-rance	Reported Accuracy/Error	Fusion/Drift Reduction Technique
Massachusetts Bay Engineering Trials	INS (iXblue PHINS C7) + DVL	~ 10 km	20–30 min (short tests)	Average drift ≈ 0.1% (~10 m error over 10 km)	INS/DVL fusion using EKF/UKF
Under-Ice Mission in the Beaufort Sea (ICEX20)	INS + DVL + Acoustic (HydroMAN-ICNN)	~ 11 km	~2.5 h (typical under-ice run)	Majority of corrections < 15 m; 77% < 20 m error	Multi-sensor fusion with acoustic updates (LBL/USBL aiding)
SWARMs (European Project)	INS, DVL, Acoustic, GPS (when surfaced)	10 km² area	~24 h	Position error within 15–20 m	Decentralised sensor fusion and adaptive navigation algorithms
AQUASwarm (European Project)	INS, DVL, Acoustic, Environmental sensors	3–5 km coastal	~20 h	Absolute navigation error ≈ 15 m	Real-time adaptive mission planning with sensor fusion
Autosub3 (Antarctica Deployment)	INS + DVL + USBL	Up to 60 km	60–70 h	Corrected drift error < 20 m over 60 km (with USBL updates)	INS/USBL fusion for long-range drift correction
Sentry (WHOI AUV)	INS, DVL, USBL/USBL, Pressure, Optical	100–150 km (estimated)	(In days)	Heading: 0.5°; Depth: 2 cm; Position error target < 20 m when aided by surface fixes	Fusion of high-grade INS with acoustic updates and periodic surfacing

TABLE 7.4
Characteristics of Various Positioning Systems for UUVs

Positioning System	Advantages	Disadvantages
Inertial navigation system (INS)	-No need for external signals -Can operate in GPS-denied environments	-Drift over time leading to decreased accuracy -Requires periodic calibration
Long baseline (LBL)	-High positional accuracy -Effective in deep water -Suitable for complex environments	-Requires installation of multiple transponders -Limited mobility during setup
Short baseline (SBL)	-Less complex than LBL -Suitable for shallower waters -Easier and quicker to deploy	-Lower accuracy compared to LBL -Limited range and precision
Ultra-short baseline (USBL)	-Real-time tracking -Requires only a single transducer -Good for dynamic environments	-Less accurate at greater distances -Performance can be affected by surface conditions
Doppler velocity log (DVL)	-Provides velocity relative to the seafloor or water column -Enhances positioning accuracy if combined with other systems	-Accuracy affected by water and seafloor characteristics -Not a standalone positioning system
Dead-reckoning	-Simple and cost-effective -Useful for estimating position in the absence of other data	-Accuracy degrades over time -Dependent on accurate initial conditions
Acoustic modem systems	-Enables communication between UUVs and surface vessels -Can assist in positioning and data transfer	-Limited bandwidth -Communication can be affected by water conditions and distance
LiDAR (light detection & ranging)	-Provides high-resolution 3D data -Effective for shallow-water mapping	-Limited depth penetration -Requires clear water conditions

from both experimental and CFD studies, form the foundation of precise navigation and control. Through systematic estimation of hydrodynamic coefficients, incorporation of fin-hull interference effects, and time-domain simulation of motion equations, one can accurately predict and optimise UUV manoeuvrability under various operational scenarios. Such a model, validated by both theory and experimental data, provides the digital 'twin' of the real vehicle.

While the current state-of-the-art allows for navigation accuracies that are sufficient for many scientific and operational missions, challenges remain in ensuring robustness in highly dynamic and unpredictable environments. Improvements in these areas are critical for the success of scientific, industrial, and military underwater missions, particularly to enable extended duration missions underwater and for collaborative (swarm) operations.

Managing environmental variability, ensuring computational efficiency, and integrating emerging sensor technologies continue to be highly challenging. Ongoing research into adaptive filtering, real-time processing, and the integration of emerging sensor technologies promises to further reduce drift and improve the overall accuracy and reliability of underwater navigation systems.

REFERENCES

Alexandris, C., Papageorgas, P., & Piromalis, D. (2024). Positioning Systems for Unmanned Underwater Vehicles: A Comprehensive Review. *Applied Sciences, 14*(21). https://doi.org/10.3390/app14219671

Asai, T., Kojima, J., Asakawa, K., & Iso, T. (2000). Inspection of submarine cable of over 400 km by AUV. In *Proceedings of the 2000 International Symposium on Underwater Technology (Cat. No.00EX418)* (pp. 133–135). Tokyo, Japan: IEEE. https://doi.org/10.1109/ut.2000.852529

Bachmayer, R., Antonelli, E., Fossen, T., Indiveri, G., Yuh, J., & Ødegaard, A. (2008). *Springer Handbook of Robotics*. Heidelberg: Springer.

Bhatt, E. C., Viquez, O., & Schmidt, H. (2022). Under-ice Acoustic Navigation Using Real-time Model-aided Range Estimation. *Journal of the Acoustical Society of America, 151*(4), 2656–2671. https://doi.org/10.1121/10.0010260

Caccia, M. (2006). Vision-based Linear Motion Estimation for Unmanned Underwater Vehicles. In *2003 IEEE International Conference on Robotics and Automation (Cat. No.03CH37422).* (Vol. 1, pp. 977–982). IEEE. https://doi.org/10.1109/robot.2003.1241719

Chaudhary, M., Goyal, N., Benslimane, A., Awasthi, L., Alwadain, A., & Singh, A. (2023). Underwater Wireless Sensor Networks: Enabling Technologies for Node Deployment and Data Collection Challenges. . *IEEE Internet Things Journal, 10*, 3500–3524.

De Barros, E., Pascoal, A., & De Sa, E. (2004, July). AUV Dynamics: Modelling and Parameter Estimation Using Analytical, Semi-empirical, and CFD Methods. *IFAC Proceedings, 37*, 369–376. https://doi.org/10.1016/s1474-6670(17)31760-3

Eustice, R. M., Singh, H., Leonard, J. J., Walter, M. R., & Ballard, R. D. (2005). Visually Navigating the RMS Titanic with SLAM Information Filters. In *Robotics: Science and Systems I* (pp. 451–460). Cambridge, MA: MIT Press.

Fossen, T. (1991). *Modeling and Control of Underwater Vehicles*. Trondheim, Norway: Marine Cybernetics.

González-García, J., Gómez-Espinosa, A., Cuan-Urquizo, E., García-Valdovinos, L. G., Salgado-Jiménez, T., & Cabello, J. A. (2020). Autonomous Underwater Vehicles: Localization, Navigation, and Communication for Collaborative Missions. *Applied Sciences, 10*(4). https://doi.org/10.3390/app10041256

Hegrenæs, Ø., Ramstad, A., Pedersen, T., & Velasco, D. (2016). Validation of a New Generation DVL for Underwater Vehicle Navigation. In *IEEE/OES Autonomous Underwater Vehicles (AUV)* (pp. 342–348). Tokyo: IEEE. https://doi.org/10.1109/AUV.2016.7778694

Heshmat, M., Saoud, L. S., Abujabal, M., Sultan, A., Elmezain, M., Seneviratne, L., & Hussain, I. (2025). Underwater SLAM Meets Deep Learning: Challenges, Multi-Sensor Integration, and Future Directions. *Sensors, 25*(11). https://doi.org/10.3390/s25113258

Hong, L., Wang, X., & Zhang, D.-S. (2024, August). CFD-Based Hydrodynamic Performance Investigation of Autonomous Underwater Vehicles: A Survey. *Ocean Engineering, 305.* https://doi.org/10.1016/j.oceaneng.2024.117911

Kinsey, J. C., Eustice, R. M., & Whitcomb, L. L. (2006). *A Survey of Underwater Vehicle Navigation: Recent Advances and New Challenges.* IFAC Conference of Manoeuvring and Control of Marine Craft. Lisbon.

Kongsberg Discovery. (2024, September 4). Smashing AUV Records, HUGIN Endurance Completes a Multi-week Fully Autonomous Mission. *Kongsberg.com.* Retrieved May 2025, from https://www.kongsberg.com/newsroom/news-archive/2024/smashing-auv-records-hugin-endurance-completes-a-multi-week-fully-autonomous-mission/

Kongsberg Maritime. (2023). *HiPAP and cNODE Positioning Systems.*

Leonard, J., & Bahr, A. (2016). Autonomous Underwater Vehicle Navigation. In M. R. Dhanak & N. I. Xiros (Eds.), *Springer Handbook of Ocean Engineering* (pp. 341–358). Springer.

Liu, Y., & Bucknall, R. (2019). *Navigation and Control of Autonomous Marine Vehicles.* The Institution of Engineering and Technology. https://doi.org/10.1049/pbce019e

McPhail, S. e. (2009). *Exploring Beneath the Ice Shelf with Autosub3.* OCEANS 2009 Europe.

Meduna, Z., Petillot, Y., & Lane, D. (2008). Long-term Autonomous Underwater Vehicle Navigation Using Terrain Matching. *IEEE Journal of Oceanic Engineering, 33,* 103–121.

Paull, L., Saeedi, S., Seto, M., & Li, H. (2014). AUV Navigation and Localization: A Review. *IEEE Journal of Oceanic Engineering, 39*(1), 131–149.

Randeni, S., Schneider, T., Bhatt, E. C., Víquez, O. A., & Schmidt, H. (2022, November). A High-resolution AUV Navigation Framework with Integrated Communication and Tracking for Under-ice Deployments. *Journal of Field Robotics.* https://doi.org/10.1002/rob.22133

Ray, A., Sen, D., Singh, S. N., & Seshadri, V. (2016). *Manoeuvrability Assessment of Underwater Vehicles by Semi-Empirical, Numerical and Experimental Studies.* 31st Symposium on Naval Hydrodynamics. ONR, Monterey, CA.

Teixeira, F. C., & Pascoal, A. M. (2007). Geophysical Navigation of Autonomous Underwater Vehicles. *IFAC Proceedings, 40*(17), 117–122. https://doi.org/10.3182/20070919-3-HR-3904.00022

Tiano, A., & Ferri, G. (2018). Underwater Vehicle Navigation Using SINS, DVL and LBL with Intelligent Sensor Data Fusion. *Journal of Intelligent & Robotic Systems, 91,* 263–278.

Yu, Z., Zhang, Q., Chen, G., Huang, C., & Lan, J. (2023). *Submarine Geophysical Navigation of Autonomous Underwater Vehicle Using Extended Kalman Filter.* 2023 IEEE International Conference on Electrical, Automation and Computer Engineering (ICEACE). Changchun, China. https://doi.org/10.1109/ICEACE60673.2023.10442118

Yuan, X., Martínez, J.-F., Pons, J. S., & Eckert, M. (2017, May). AEKF-SLAM: A New Algorithm for Robotic Underwater Navigation. *Sensors.* https://doi.org/https://doi.org/10.3390/s17051174

Zhang, S., Zhao, S., An, D., Liu, J., Wang, H., Feng, Y., . . . Zhao, R. (2022, November). Visual SLAM for underwater vehicles: A survey. *Computer Science Review, 46.* https://doi.org/10.1016/j.cosrev.2022.100510

8 Autonomous System Architecture

8.1 AUTONOMY AND ITS LEVELS

Rather than looking at UUVs as a collection of physical components (equipment, sensors, hardware, software), a more nuanced perspective looks at them as a system, with conceptual 'layers' based on functionality. Such a perspective is widely adopted for describing networks, emphasising the need to connect, exchange information, and respond to stimuli.

The functionality of any UUV depends critically on a well-structured control architecture, which governs the interaction between software, hardware, and communication subsystems. This architecture determines how commands are executed, how sensor data is processed, how navigation and autonomy are achieved, and how modules interoperate, particularly in distributed and modular AUV platforms (Antonelli, 2018).

A widely accepted framework (Huang, 2008) categorises autonomy into levels as listed in Table 8.1. Representative ROV/AUV models are listed along with their demonstrated autonomy level and key capabilities.

8.2 LAYERS OF SYSTEM ARCHITECTURE

UUVs (AUVs and ROVs) can be considered as made up of distinct conceptual layers for hardware, software, and control. These layers work together to enable navigation, data acquisition, and mission execution in underwater environments.

8.2.1 Hardware Layer

For subsea systems (UUVs), this includes the vehicle itself, with its propulsion system (thrusters, propellers, or buoyancy engines), sensors (pressure, depth, temperature, etc.), and on-board computers.

In case of ROVs, we have systems outside the water (Topside) and the connection to the ROV (Umbilical). The Topside systems (for ROVs) consist of the components on the surface vessel, such as the control console, power supply, and communication equipment. The umbilical cable is the physical link that transmits/receives data and transmits power from the surface to the underwater vehicle.

DOI: 10.1201/9781003737476-8

TABLE 8.1
Levels of Autonomy and Examples of UUVs

Autonomy Level	Description	Examples of UUVs	Key Capabilities
Level 0	Remotely operated vehicles (ROVs) with direct human control	WHOI MicroROV, VideoRay Pro 4	Direct teleoperation, real-time video feedback, limited on-board processing
Level 1	Basic autonomous vehicles executing predefined mission scripts without adaptive re-planning	Early generation AUVs (e.g., early REMUS prototypes)	Fixed mission execution, simple depth/heading control, basic sensor feedback
Level 2	Semi-autonomous vehicles with adaptive control and environment reaction capability	REMUS 600/100, Bluefin-21 Glider, SeaExplorer	Adaptive navigation, obstacle avoidance, intermittent operator updates, fault tolerance
Level 3	Fully autonomous systems with on-board high-level decision-making and mission re-planning capabilities	Advanced prototypes (e.g., the NEXT AUV series, some DARPA systems)	Real-time path re-planning, multi-sensor fusion, autonomous obstacle avoidance, self-recovery

8.2.2 Software Layer

- **Operating System**: A real-time operating system manages the vehicle's resources and provides a platform for other software components.
- **Middleware**: This layer facilitates communication between different software modules and hardware components.
- **Mission-Specific Software**: This includes modules for navigation, path planning, obstacle avoidance, data acquisition, and communication protocols.
- **Sensor Data Processing**: Software algorithms process raw sensor data to provide usable information for navigation and other tasks.
- **Control Algorithms**: These algorithms translate desired vehicle movements into commands for the actuators.

8.2.3 Control Layer

- **Guidance System**: This layer determines the vehicle's overall path and objectives, often involving path planning and behaviour selection.
- **Control System**: This layer implements the control algorithms to steer the vehicle, maintain depth, and execute manoeuvres.
- **Fault Detection and Handling**: The system monitors for errors and implements contingency plans to ensure mission safety.

8.2.4 Task Layer

- **Perception**: This involves using sensors to understand the environment and detect obstacles or features of interest.
- **Communication**: This handles data transmission between the vehicle and the surface (ROV) or other vehicles (AUV).
- **Motion Control**: This layer translates high-level commands into specific movements for the vehicle's actuators.
- **Fault Treatment**: This layer handles any errors or malfunctions that may occur during the mission.

8.2.5 Alternative Representation

AUV control architecture can be represented in terms of hierarchical layers in more than one way (Caccia, 2006). Another variant of the layers mentioned above (8.2.1 to 8.2.4) is given below. Here, all layers refer to software or communication protocols for different aspects of AUV operation (Yoerger et al., 2007).

- Mission Layer:
  - Responsible for high-level mission planning and decision-making.
  - Implements logic for mission execution, goal prioritisation, and contingency handling.
  - May use behaviour trees, finite state machines, or AI-based planners.
- Task Execution Layer:
  - Decomposes mission goals into executable tasks (e.g., waypoint following, data collection).
  - Coordinates sensor payloads, control routines, and subsystems in response to mission needs.
- Vehicle Control Layer:
  - Executes control laws (e.g., PID, adaptive control) for guidance, navigation, and control (GNC).
  - Interfaces with the vehicle's actuation systems (thrusters, fins).
- Device and Hardware Abstraction Layer:
  - Manages direct control over hardware components such as sensors, communication modules, and actuators.
  - Provides hardware abstraction to ensure modularity and reusability.

8.3 PROTOCOL STACK

A Protocol Stack is a layered model showing how data flows from a high-level application (like mission control software) down to the physical communication hardware (e.g., an acoustic modem), and vice versa.

Protocol Layer placement in an AUV/ROV stack is shown below, with examples:
[High-level Autonomy & Control]

└── JAUS, ROS, MOOS-IvP, DDS, IMC
[Transport & Messaging]
└── DDS-RTPS, SOME/IP, TCP/UDP, NMEA
[Physical Communication Interfaces]
└── RS-232, RS-485, CAN, I²C, SPI
[Specialised Underwater Comms]
└── JANUS (acoustic), WHOI Micromodem, BlueComm (optical)

A typical protocol stack in the context of AUV communication is shown below.

High-Level Software
↕
Middleware/Control Protocols (e.g., MOOS, ROS, DDS)
↕
Transport Layer (e.g., TCP/UDP)
↕
Acoustic Protocol Layer (e.g., S2C LLP, WHOI micromodem packet format)
↕
Modem Hardware Interface (RS-232, Ethernet, etc.)
↕
Physical Layer (Sound propagation in water)

Control and communication protocols used in AUVs/ROVs extend across different layers of the system architecture, as shown in Table 8.2. It is essential to know which protocol(s) are supported by particular equipment, in order to determine whether (or how) it can be integrated into the UUV being designed or developed.

8.4 AUV HARDWARE ARCHITECTURE

AUVs must operate in complex, dynamic underwater environments, which demand that all components—from the hull and propulsion system to the sensor suite and onboard computers—are tightly integrated. The design must satisfy several conflicting requirements: high energy efficiency, robust operation under high pressure, low maintenance, and high autonomy. Holistic design approaches integrate these diverse subsystems using common design languages and modelling techniques.

TABLE 8.2
Protocols Used at Different Layers of System Architecture

Layer	Examples	Function
Application layer (high-level)	JAUS, ROS, DDS, MOOS-IvP	Autonomy, control logic, inter-module messaging
Transport/messaging layer	TCP/UDP, DDS-RTPS	Message encoding, delivery across subsystems
Hardware/bus communication layer (low-level)	SPI, I²C, CAN, RS-232	Communication between sensors, controllers, and low-level boards

8.4.1 Hardware Architecture and Data Protocols

One of the key trends in AUV development is modularity. Modular hardware architecture divides the vehicle into discrete, functionally independent 'boxes' or modules (e.g., power module, propulsion module, sensor module, communication module). Each module is designed with well-defined interfaces for power, data, and mechanical connections. Modularity offers several advantages:

- **Flexibility**: Modules can be upgraded or replaced independently, allowing the AUV to be reconfigured for different missions.
- **Scalability**: New functionalities can be added by integrating additional modules without redesigning the entire system.
- **Maintainability**: Fault isolation is simplified when modules operate as separate entities, and rework or repair can be done on individual modules.
- **Standardisation**: Using industry-standard connectors, communication protocols, and mechanical interfaces ensures that modules from different suppliers or developed in-house are interoperable.

For example, the ROV Pionero500 uses a modular design in which subsea components are grouped into pressure-sealed 'boxes' that handle power, computation, sensor input, and control output. This modular approach reduces the complexity of the physical architecture and facilitates rapid prototyping and testing.

8.4.2 Standardised Interfaces and Data Protocols

One of the main challenges in AUV development is ensuring that various hardware and software modules, often provided by different vendors or developed by different teams, can interoperate. Standardisation can be addressed by the following means (see Table 8.3):

- **Defining Communication Protocols**: Adopting common protocols for data exchange (such as DDS, MQTT, or ROS topics) that specify message formats, timing, and quality-of-service (QoS) parameters.
- **Interface Standardisation**: Specifying electrical and mechanical interfaces for hardware modules (e.g., connectors, power specifications, and signalling standards) to guarantee compatibility. For instance, using standardised connectors (such as MIL-SPEC wet-mateable connectors) and communication buses (such as CAN or Ethernet) facilitates integration.
- **Data Abstraction**: Using middleware that abstracts raw data into standardised structures (e.g., using a common data model for sensor outputs). This allows software modules to process data uniformly, regardless of the sensor manufacturer.
- **Use of Open Standards**: Leveraging open-source platforms (e.g., the Robot Operating System (ROS)) which provide well-documented standards for message passing, service definitions, and system configuration. Open standards ensure that new components are easier to integrate and that the overall system remains scalable and maintainable.

TABLE 8.3
Overview of Standard Interfaces and Data Protocols in AUV Software

Layer/Interface	Typical Protocol/ Standard	Description	Advantages
Hardware abstraction	RS-232/RS-422, I^2C, SPI	Standard electrical communication for sensors and actuators	Low-level simplicity; well-established hardware interfaces
Middleware	ROS, MOOS-IvP, DDS	Publish/subscribe messaging; standardised APIs for data exchange	Decouples modules; supports distributed processing
Data formats	JSON, XML, Protocol Buffers	Common formats for representing sensor data and messages	Facilitates interoperability; human-readable if needed
Networking	Ethernet, CAN Bus	High-speed, reliable data exchange between modules	Supports high data rates; robust against interference
Real-time communication	SOME/IP, DDS QoS	Ensures timely delivery of control commands and sensor data	Guarantees low latency; supports critical mission functions

8.5 AUV SOFTWARE ARCHITECTURE

The software architecture for an AUV must support real-time operation, concurrent processing, and flexible mission management. In this section, the main components of modular software architecture for AUVs are discussed.

8.5.1 Layered Architecture

A common approach is to adopt a layered model that separates concerns and isolates functionality. A typical AUV software architecture might include the following layers:

Device and Hardware Abstraction Layer:

- Provides drivers and interfaces to sensors, actuators, and communication hardware.
- Abstracts the low-level hardware details so that higher layers can operate without concern for specifics such as serial protocols or analogue-to-digital conversion.

Middleware/Communication Layer:

- Implements standard data exchange protocols (e.g., publish/subscribe, remote procedure calls).

- Often built on open standards (such as DDS or ROS topics) to facilitate interoperability and distributed processing.
- Provides a uniform messaging interface to support modularity and reusability.

Control and Decision Layer (Backseat Driver):

- Contains high-level control algorithms, mission planning, and decision-making logic.
- Utilises models such as behaviour trees or state machines to structure complex autonomous behaviour.
- Implements fault tolerance measures by monitoring subsystem status and re-planning missions if necessary.

Application/User Interface Layer:

- Provides interfaces for mission configuration, monitoring, and control by human operators.
- Often includes visualisation tools and diagnostic displays.
- Facilitates user interaction through standard graphical user interfaces (GUIs) or web-based dashboards.

Data Management and Logging Layer:

- Manages data storage, retrieval, and synchronisation between on-board systems and surface stations.
- Uses standardised data protocols (e.g., JSON, XML, or binary formats) to ensure that data can be processed by third-party tools.
- Implements real-time logging and fault reporting mechanisms.

8.5.2 Mission Computer Modularity

The mission computer serves as the 'brain' of the AUV and is responsible for coordinating all subsystems. Modularity in the mission computer design is achieved by:

Partitioning Software Components: Separating mission-critical functions (such as navigation and control) from auxiliary functions (such as data logging and sensor calibration). This partitioning is often implemented via a microkernel or a modular real-time operating system (RTOS).

Standardised Interfaces: Using standardised APIs (application programming interfaces) for communication between modules. This enables components developed by different teams or sourced from different vendors to interact seamlessly.

Plug-and-Play Capability: Designing the system so that new modules (e.g., a new sensor driver or control algorithm) can be integrated without affecting existing functionality.

Fault Isolation: Structuring the mission computer so that if one module fails, it does not cascade and cause a total system shutdown. This is achieved by isolating processes (using, for example, containerisation or sandboxing techniques).

8.5.3 Backseat Driver Philosophy

The 'backseat driver' concept is a design philosophy wherein a high-level module monitors the system's performance and external conditions, and issues strategic commands to the low-level control module (the 'frontseat driver'). This approach has several advantages:

Decoupling: It separates high-level decision making from low-level control, allowing each to be developed and tested independently.

Autonomy and Adaptability: The backseat driver can analyse sensor data, assess environmental changes, and adjust mission plans in real time.

Safety and Fault Tolerance: By continuously monitoring the state of the system, the backseat driver can detect anomalies or faults early and take corrective action—such as rerouting the vehicle or activating redundancy protocols—thus enhancing overall system robustness.

Ease of Integration: New decision-making algorithms and behaviours can be integrated into the backseat driver module without impacting the critical real-time control loops.

Frameworks such as MOOS-IvP have been used to implement this philosophy in underwater robotics. In MOOS-IvP, the MOOS middleware handles message passing between modules while IvP Helm provides autonomy through behaviour-based decision-making. This separation of concerns illustrates the backseat driver approach in practice.

8.6 FAULT TOLERANCE STRATEGIES

Fault tolerance is essential for AUVs, given the difficulty and expense of retrieving a vehicle that has failed underwater. Strategies for achieving fault tolerance include the following:

Redundancy: Incorporating duplicate sensors, communication links, and processing units so that the failure of one component does not compromise the entire system. For example, a redundant thruster design can ensure that if one thruster fails, the vehicle can still maintain control.

Graceful Degradation: Designing systems to reduce performance gradually rather than failing catastrophically. If a non-critical module fails, the system can still operate in a degraded mode until recovery is possible.

Self-Diagnostics and Reconfiguration: Implementing health monitoring algorithms that continuously assess the status of each module. When a fault is detected, the system can reconfigure itself (e.g., by switching to backup modules or rerouting data flows) to maintain safe operation.

Modular Isolation: Using techniques such as process isolation and sandboxing to prevent a fault in one software module from propagating to others.

Robust Communication Protocols: Ensuring that data protocols include error detection, correction, and retransmission mechanisms to handle transient communication failures.

8.6.1 Object-Oriented Programming and the Actor Model

Modern AUV software often relies on object-oriented programming (OOP) to create reusable, modular code. OOP promotes the creation of abstract classes for common functionality (e.g., a generic 'Instrument' class for sensors), which can then be extended for specific hardware. This approach reduces code duplication and improves maintainability.

Additionally, many AUV control systems have adopted the actor model for managing concurrency. In this model, independent 'actors' (software objects) process data and communicate via asynchronous message passing. This design greatly simplifies the handling of multiple concurrent tasks (such as sensor data acquisition, control loop execution, and mission planning) and avoids common pitfalls such as deadlocks and race conditions.

8.7 MODULAR AND INTEROPERABLE ARCHITECTURES

As AUV platforms become more modular and mission-flexible, there is a growing need for standardised communication and control interfaces. This ensures that new subsystems—such as sonar modules, autonomy engines, or navigation sensors—can be integrated with minimal effort. This is where 'middleware' and interoperability standards come into play.

8.7.1 Middleware

A robotics 'middleware' can essentially be thought of as a software layer, composed of several modular packages, which has the task of collecting and harmonising the information coming from the on-board sensors, making them available to the various processing nodes. These nodes must in turn process this data according to their specific function (tracking, communications, etc.) and pass the outputs to other nodes or directly to sensors and/or actuators of the robotic system. At the end of the various elaborations, the middleware will be responsible for the passage of the system into another state, e.g., the execution of a specific action (Costanzi, Fenucci, & Manzari, 2020).

Some significant and common standards for control architecture are listed below:

- MOOS-IvP (Mission Oriented Operating Suite—Interval Programming.
- ROS (Robot Operating System).
- JAUS (Joint Architecture for Unmanned Systems).
- DDS (Data Distribution Service).

The current de facto standard middlewares for UUVs include MOOS and ROS, which are introduced here.

8.7.2 MOOS-IvP

MOOS is a lightweight middleware framework developed by the Oxford Mobile Robotics Group, designed specifically for marine robotics. It is optimised for mission control, real-time coordination, and long-duration, autonomous operations. Both MOOS and ROS are publish-and-subscribe systems, which provide the communication of arbitrary data throughout a network (Benjamin et al., 2010).

However, in order to complete mission objectives, a robotic system also requires a deliberative component in addition to its reactive aspects (e.g., avoiding obstacles). This is the objective of the MOOS process 'IvP Helm', which acts as an autonomous decision-making engine to execute in the 'backseat' of the robotic platform. MOOS-IvP stands for 'Mission Oriented Operating Suite – Interval Programming'. MOOS-IvP is the extended version that includes IvP Helm, a behaviour-based decision engine used for high-level autonomy.

MOOS is the actual robotics middleware, abstracting a TCP/IP-based interprocess communication (IPC) protocol, while IvP is a multi-objective optimisation framework for constructing complex autonomous behaviours from a collection of simpler routines. The whole environment is composed of several other processes, which communicate using the MOOSDB database as a broker. The rate of interactions across information channels, called 'topics', is controlled tightly by the MOOSDB, occurring synchronously at a predefined rate. The exchange of information occurs between TCP ports, which may or may not exist on the same physical machine (Benjamin et al., 2010).

Therefore, MOOSDB is relatively rigid, and each topic is bound to a startup-defined data type, with induced additive latencies because of the travel path of information (node-database-node) (Costanzi, Fenucci, & Manzari, 2020). On the other hand, MOOS-IvP is ideal for deploying in bandwidth-limited marine environments, or for behaviour-based mission execution.

8.7.3 ROS

ROS (Robot Operating System) is a flexible, modular middleware framework for developing robot software. It is not an actual operating system, but a collection of tools, libraries, and conventions that simplify the task of building complex and scalable robot systems. Originally developed for ground robots, ROS is now widely used in marine robotics, especially for rapid prototyping, data handling, and integration of heterogeneous subsystems (DeMarco, West, & Collins, 2011).

In ROS, data are transferred using peer-to-peer communications (on the other hand, all data within a MOOS system are transmitted through the MOOSDB). The basic unit of interaction in ROS is a 'message', which is typically exchanged on a 'topic': from a node's perspective, messages are published synchronously and read from a subscription asynchronously. In ROS, there are two extensions to the typical publish/subscribe mechanism: 'services' and 'actions'. A 'service' simply takes an input and returns an output, as a typical service does in computer science literature. An 'action' keeps an internal state on a longer time scale because it is called with a goal, emits feedback during the action, and finally returns a single result at the end of the action. An important advantage of ROS over MOOS is its capability of handling very large datasets, e.g., live high-resolution video and multidimensional point clouds from LiDAR-like devices.

Tools available in ROS include visualisation (RViz), simulation (Gazebo), logging, diagnostics, and bagfile recording. It has an extensive library including SLAM, computer vision, path planning, and control algorithms. Its subsequent version, ROS 2, adds support for real-time, DDS-based communication and better multi-robot systems.

8.7.4 DDS

DDS (Data Distribution Service) is a standardised middleware protocol and API (defined by the OMG—Object Management Group) for real-time, scalable, high-performance communication between distributed systems.

Unlike ROS and MOOS, which are frameworks with many tools and concepts, DDS is purely a communication layer like a 'transport backbone', designed to: Deliver real-time, low-latency, and reliable data exchange; Enable decentralised, peer-to-peer communication; and Support Quality of Service (QoS) controls (latency, reliability, history, deadlines, etc.). ROS2 is built natively on DDS (while ROS 1 does not use DDS by default). MOOS does not use DDS, and uses its own internal centralised bus (MOOSDB), where all components publish/subscribe to named variables.

8.7.5 JAUS in AUV Control Architectures

JAUS (Joint Architecture for Unmanned Systems) is a specification introduced by the US DoD to provide a basis for interoperability between unmanned systems (Whitsitt & Sprinkle, 2011). Developed under SAE AS-4 standards, JAUS is a message-based, platform-independent standard for the command and control of unmanned systems, especially in military/defence AUVs. It defines a set of standard interfaces, messages, and component behaviour to ensure interoperability across heterogeneous unmanned systems—land, air, or underwater.

A set of standard application layer interfaces called 'JAUS Mobility Services' has been defined. JAUS Services provide the means for software entities in an unmanned system (or system of unmanned systems) to communicate and coordinate their activities. The Mobility Services represent vehicle platform-independent capabilities commonly found across all domains and types of unmanned systems (referred to as UXVs). At present, over 15 services are defined in this document, many of which have been updated to support Unmanned Underwater Vehicles (UUVs) (JAUS Mobility Service Set AS6009A, 2023).

Based on JAUS standards, Software Development Kits (e.g., C++ based middleware toolkit) are available commercially.[1] Such middleware can be used in an AUV so as to standardise its software in order to make it interoperable with other SAE JAUS-based systems. Open source options are also available.[2]

In the context of AUVs, JAUS has the following features:

- **Interoperability and Modularity**: JAUS promotes a component-based architecture, where subsystems (e.g., navigation, propulsion, sensor management) are represented as software components with well-defined interfaces. This allows AUV developers to integrate components from multiple vendors, replace or upgrade subsystems without affecting the rest of the system, and simplify validation and testing via defined interface behaviour (Costanzi, Fenucci, & Manzari, 2020).
- **Distributed Communication**: JAUS supports peer-to-peer messaging, enabling distributed control. In an AUV, this allows the autonomy engine to directly communicate with the propulsion system, mission management systems to dynamically reconfigure payloads, and sensor fusion nodes to interact seamlessly with localisation systems.
- **Multi-vehicle Coordination**: When operating as part of a fleet (e.g., swarm missions), JAUS supports coordinated behaviour by standardising messages for position updates, task delegation, health monitoring and status reporting.
- **Safety and Diagnostics**: JAUS includes message sets for health monitoring, fault reporting, and emergency control. These are essential for underwater vehicles that may be deployed for extended periods in unstructured environments.

8.7.6 Comparison of Middleware

The development of MOOS-IvP in the research community is continuous and ongoing, but its usage in industry is very limited. This may be a consequence of the lack of insurance on backward compatibility, which is based on community agreement and not on an enforced standards. While MOOS has historically been popular within the underwater robotics community, ROS is now by far more pervasive in a multi-domain context (ground, sea, air). The main reasons for this are that ROS is reconfigurable and relatively easy to use. There are bindings for both C++ and Python, and it is regularly updated (at least yearly). There are no rigid standards guiding development, but it is so widely used that there is extreme prejudice against breaking

TABLE 8.4
Summarised Comparison of ROS and MOOS Middleware

Feature	ROS	MOOS (MOOS-IvP)
Design origin	Ground/mobile robotics (general)	Marine robotics, mission control
Communication	Pub-Sub via topics, services	Central database with variable updates
Autonomy engine	Custom (or via Behaviour Trees/AI)	IvP Helm (multi-objective behaviour engine)
Real-time support	ROS 1: No; ROS 2: Yes (via DDS)	Yes (simple, deterministic message system)
Ease of use	Rich libraries, tools, rapid dev	Lightweight, but requires C++ expertise
Field proven in	Prototypes, commercial AUVs, drones	Long-range AUVs, WHOI missions, swarms

backward compatibility (Costanzi, Fenucci, & Manzari, 2020). The main features of MOOS-IvP and ROS are compared in Table 8.4.

The main features of MOSS, ROS, DDS and JAUS are compared in Table 8.5.

8.8 LINKING CONCEPTS

8.8.1 GNC and Middleware

The middleware frameworks (ROS, MOOS, JAUS, and DDS) that we discussed connect the guidance/control algorithms with: sensor data inputs (e.g., position, velocity, orientation); mission/task commands (e.g., new waypoint, change depth); and actuator interfaces (e.g., thruster setpoints). The summary of the roles of GNC and middleware would help in clarifying these concepts (see Table 8.6).

Thus, Middleware enables the GNC system to be modular and distributed. Control architecture organises GNC within a layered structure, maintaining separation of concerns. JAUS, ROS, MOOS, and DDS all serve to support the communication and orchestration of GNC algorithms, but at different layers and with different philosophies.

Control strategies are central to GNC, but also extend into autonomy, coordination, and mission execution. Low-level control stabilises and moves the vehicle. High-level control makes decisions and governs complex behaviours. Middleware like ROS, MOOS, and JAUS enables the communication, distribution, and modular deployment of these strategies. Control strategies are not limited to GNC, but GNC is the most mathematically intensive and tightly coupled domain for them.

8.8.2 Modularity in Hardware and Software

To illustrate the key concepts described above, Table 8.7 summarises various aspects of AUV design and software architecture.

TABLE 8.5
Comparison of Popular Middleware Protocols and Frameworks

Feature	DDS	ROS (1 & 2)	MOOS	JAUS
Description	Middleware protocol	Robot software framework	Lightweight middleware + autonomy framework	Interoperability and messaging standard
Design origin	Real-time distributed systems (OMG)	Ground/mobile robotics (Willow Garage, OSRF)	Marine robotics (Oxford MRG, WHOI)	DoD/SAE standard for unmanned systems
Communication model	Peer-to-peer, publish/subscribe, QoS-controlled	ROS 1: Centralised pub-sub ROS 2: Peer-to-peer via DDS	Centralised variable store (MOOSDB)	Peer-to-peer message passing (strict message formats)
Standardisation	Yes (OMG DDS)	ROS 1: Community-based ROS 2: OMG DDS-compliant	No formal standard (de facto in marine research)	Yes (SAE AS-4 series standards)
Real-time support	Yes, deterministic with QoS	ROS 1: No ROS 2: Yes (through DDS)	Yes (simple, predictable scheduling)	Yes, deterministic messaging
Scope	Transport layer only	Full robotics stack (control, planning, perception)	Mission-level control, autonomy behaviours	Messaging structure, service definitions for unmanned systems
Modularity/ interoperability	Moderate (depends on DDS vendor)	ROS 2: High ROS 1: Moderate	Moderate (tight coupling to MOOSDB)	High (strong focus on cross-vendor/ component interfaces)
Autonomy support	Not included (use with higher layers)	Yes (via planners, behaviour trees, AI integration)	Yes (IvP Helm behaviour engine)	Limited (focus on control, not behaviour logic)
Ease of use	Complex to configure	ROS 1: Easy ROS 2: Moderate (more robust)	Moderate (C++ based, compact APIs)	Steep learning curve, rigid specification
Field proven applications	Defence systems, ROS 2 backends, DDS-native systems	Prototypes, AUVs, drones, industrial robotics	WHOI AUVs (REMUS), academic fleets	US DoD AUV/UUV programs, Kingfish, Knifefish, etc.
Primary strengths	Deterministic, decentralised, scalable	Rich ecosystem, modularity, perception support	Lightweight, reliable, mission-focused	Standardised, interoperable control across vendors
Limitations	No autonomy tools, requires DDS vendor/ runtime	ROS 1 not real-time; ROS 2 configuration is complex	Less scalable, no advanced perception support	Rigid message format, limited dynamic behaviour execution

TABLE 8.6
Summary of Roles of GNC Module and Their Interaction through Middleware

Component	Role in GNC	Interaction via Middleware
Guidance module	Plans path, selects desired heading/ speed	Receives waypoints, sends goals to the control module
Control module	Executes motion using control laws (e.g., PID)	Converts goals to actuator signals
Navigation module	Estimates position from sensors (IMU, DVL, sonar)	Publishes pose for use by guidance/ control
Middleware	Connects GNC modules to sensors, actuators, and the autonomy engine	ROS: topics/services MOOS: MOOSDB vars JAUS: standard messages

TABLE 8.7
Comparison of Modular Hardware and Software Architecture Features

Feature	Hardware Modularity	Software Modularity	Benefits
Functional partitioning	Discrete 'boxes' (power, CPU, sensor, communication)	Layered architecture (device, middleware, control, UI)	Simplifies integration; allows independent upgrade
Standardised interfaces	Use of industry-standard connectors, power interfaces	Common data protocols (ROS topics, DDS, JSON/ XML)	Ensures interoperability across modules
Fault tolerance	Redundant components; compartmentalised failure isolation	Exception handling; modular isolation; actor model	Enhances reliability; allows graceful degradation
Ease of reconfiguration	Plug-and-play modules that can be swapped easily	Dynamic loading of modules; service-oriented architecture	Reduces downtime; supports rapid reconfiguration
Scalability	Expandable payload bays; multi-box design	Modular, loosely coupled components; containerised apps	Facilitates integration of new technologies
Maintenance and testing	Independent testing of each module (unit testing, IV&V)	Unit testing; continuous integration; simulation frameworks	Reduces rework; enhances system robustness

8.8.3 Other Considerations

AUVs may use either 'hierarchical' or 'behaviour-based' control architectures. ROVs typically rely on a hierarchical structure, since they operate (almost) always under human control. Other than hierarchical, control architectures could be characterised as 'heterarchical', 'subsumption', or 'hybrid'.

Communication networks are a major feature of any system architecture. Robust communication is vital for ROVs (through the umbilical) and AUVs (using acoustic modems or other wireless technologies).

Sensor Fusion has been mentioned in an earlier chapter on navigation. AUVs often fuse data from multiple sensors (e.g., INS, GPS, sonar) to improve navigation accuracy.

8.9 FUTURE TRENDS

As the field of underwater robotics evolves, several emerging trends and research directions are poised to further improve the integration between human operators and AUV systems, enhance energy efficiency, and manage data more effectively (Roberts & Sutton, 2013).

8.9.1 Autonomous Decision-Making

8.9.1.1 Enhanced Autonomy and Adaptive Algorithms

Greater levels of autonomy are expected with advanced adaptive algorithms that can dynamically balance energy consumption, sensor operation, and communication needs. Advances in artificial intelligence, particularly in deep reinforcement learning and adaptive control, will enable AUVs to learn optimal behaviours for energy management and anomaly detection based on real-time environmental conditions.

8.9.1.2 Swarm and Collective Behaviours

Research on swarm robotics is rapidly progressing. The integration of multiple AUVs into a coordinated swarm that can share data and distribute tasks efficiently is a major research focus. This will require further development of human–swarm interfaces that allow a single operator to monitor and control a large fleet with minimal cognitive load. Adaptive collaborative control and distributed decision-making models will be central to these developments (González-García et al., 2020).

8.9.2 User Interface Innovations

8.9.2.1 Immersive and Augmented Reality Interfaces

Emerging technologies such as augmented reality (AR) and virtual reality (VR) offer exciting possibilities for AUV mission control interfaces. These immersive interfaces can provide operators with a three-dimensional view of the underwater environment and the AUV's position within it. By overlaying sensor data onto a virtual model of the environment, AR systems can enhance situational awareness and reduce the cognitive burden on operators.

8.9.2.2 Multimodal Interaction

Future interfaces are likely to incorporate multiple modalities of interaction, including speech recognition, gesture control, and haptic feedback. Such multimodal systems would enable operators to interact with the AUV system in a more natural manner, potentially increasing the speed and accuracy of commands while reducing workload. Integration of natural language processing (NLP) for high-level mission commands could simplify planning and executing complex tasks.

8.9.2.3 Adaptive and Context-Aware Interfaces

Interfaces that automatically adapt based on the current mission phase and operator workload are a promising area of research. These systems could change the information density, display formats, and control options based on real-time analysis of operator performance and environmental conditions. Context-aware interfaces that adjust based on energy levels, sensor reliability, and network conditions would further enhance operator control while optimising the AUV's energy use.

8.9.3 Data Management and Analytics

8.9.3.1 Edge and Cloud Integration

Future data management architectures will likely be based on a combination of edge computing and cloud analytics. On-board edge computing will handle real-time processing and data summarisation, while cloud platforms will provide the computational power necessary for in-depth analysis and long-term storage. This distributed approach will enable operators to quickly access summarised data during missions while ensuring that raw data are preserved for post-mission analysis and future research.

8.9.3.2 Big Data and Machine Learning

The application of big data frameworks and machine learning algorithms is poised to s AUV data management. With the massive amounts of data generated, tools like Hadoop, Spark, and TensorFlow can be used to analyse trends, detect anomalies, and predict future environmental conditions. Machine learning models trained on historical mission data can improve on-board decision-making processes, enabling better adaptive behaviour in response to dynamic underwater environments.

8.9.3.3 Data Fusion and Visualisation

Improved data fusion techniques will allow for the integration of data from disparate sensors into a coherent picture. Enhanced visualisation tools will be essential to present these fused datasets in an understandable manner to operators. Innovations in 3D visualisation, interactive dashboards, and real-time mapping will continue to evolve, helping operators make informed decisions based on comprehensive data representations.

8.9.3.4 Standardisation and Interoperability

Standardisation of data formats and communication protocols is essential for ensuring interoperability among systems from different vendors. Initiatives like the JAUS (Joint Architecture for Unmanned Systems) and STANAG standards help establish common frameworks for data exchange and control commands, facilitating multi-vendor and multi-system operations. Future research will continue to refine these standards and develop common middleware that supports scalable and interoperable AUV operations (Costanzi, Fenucci, & Manzari, 2020).

In conclusion, the successful integration of human–machine interfaces, energy management strategies, and robust data processing systems is vital for the next generation of AUV missions.

NOTES

1. https://openjaus.com/sdk/
2. https://github.com/jaustoolset/jaustoolset

REFERENCES

Antonelli, G. (2018). *Underwater Robots*. Springer. https://doi.org/10.1007/978-3-319-72124-8

Benjamin, M. R., Schmidt, H., Newman, P. M., & Leonard, J. J. (2010). Nested autonomy for unmanned marine vehicles with MOOS-IvP. *Journal of Field Robotics, 27*, 834–875. https://doi.org/ 10.1002/rob.20370

Caccia, M. (2006). Autonomous Surface Craft: Prototypes and Basic Research Issues. *Autonomous Robots, 25*(4), 341–357.

Costanzi, R., Fenucci, D., & Manzari, V. (2020, July). Interoperability Among Unmanned Maritime Vehicles: Review and First In-field Experimentation. *Frontiers in Robotics and AI, 7.* https://doi.org/Volume 7 - 2020 | https://doi.org/10.3389/frobt.2020.00091

DeMarco, K., West, M. E., & Collins, T. R. (2011). An Implementation of ROS on the Yellowfin Autonomous Underwater Vehicle (AUV). *OCEANS 2011 MTS/ IEEE*, (pp. 1–7). Kona. https://doi.org/ 10.23919/OCEANS.2011.6107001

González-García, J., Gómez-Espinosa, A., Cuan-Urquizo, E., García-Valdovinos, L. G., Salgado-Jiménez, T., & Cabello, J. A. (2020). Autonomous Underwater Vehicles: Localization, Navigation, and Communication for Collaborative Missions. *Applied Sciences, 10*(4). https://doi.org/10.3390/app10041256

Huang, H.-M. (2008). *Autonomy Levels for Unmanned Systems (ALFUS) Framework.* National Institute of Standards and Technology. NIST Special Publication 1011-I-2.0. Retrieved August 2024, from https://www.govinfo.gov/content/pkg/GOVPUB-C13-cbc9faa25f6d651e046c9df607d40d59/pdf/GOVPUB-C13-cbc9faa25f6d651e046c9df607d40d59.pdf

JAUS Mobility Service Set AS6009A. (2023, October 6). SAE International. https://www.sae.org/standards/content/as6009a/

Roberts, G. N., & Sutton, R. (Eds.). (2013). *Further Advances in Unmanned Marine Vehicles.* The Institution of Engineering and Technology. https://doi.org/10.1049/PBCE002E

Whitsitt, S., & Sprinkle, J. (2011). Message Modeling for the Joint Architecture for Unmanned Systems (JAUS). In *18th IEEE International Conference and Workshops on Engineering of Computer-Based Systems* (pp. 251–259). Las Vegas, NV. https://doi.org/ 10.1109/ECBS.2011.17

Yoerger, D. R., Jakuba, M. V., Bradley, A. M., & Bingham, B. (2007). Techniques for Deep Sea Near Bottom Survey Using an Autonomous Robot. *International Journal of Robotics Research (IJRR), 26*(1), 41–54.

9 Human Interfaces and Operations

AUVs are essentially sensor nodes that capture data. Data quality and data management are of crucial importance for mission effectiveness and operations of AUVs. These aspects of data management, including data storage and transmission, are covered in the third section.

AUVs are most effective when their capabilities are multiplied through collaborative operations in swarms. However, these are limited by the challenges faced in acoustic communication. Several experiments and demonstration projects for swarm operations of AUVs and underwater sensor networks are described in the next section of the chapter.

At the end of the chapter, we touch upon integration and system-level considerations, followed by future trends and research directions.

9.1 HUMAN—AUV INTERACTION

Operators interact with AUVs via mission control centres that consolidate multiple streams of sensor data—ranging from sonar images and video feeds to navigational telemetry and diagnostic information—into a coherent interface. These interfaces must support rapid decision-making and intuitive interaction, all while accounting for the physical and operational constraints of underwater environments. For instance, acoustic communications underwater are inherently low-bandwidth and latency-sensitive, which poses challenges for real-time monitoring and control.

Human operators interact with AUVs in a variety of ways depending on the level of autonomy, mission requirements, and operational environment. This section examines the modalities of interaction, the communication channels used, and the monitoring techniques that enable operators to maintain situational awareness while remotely supervising underwater missions (Moore, Bohm, & Jensen, 2010).

9.1.1 Modes of Interaction

9.1.1.1 Direct Teleoperation

In early AUV operations, human operators were required to control the vehicle directly. This mode—often referred to as teleoperation—involves continuous manual control of the AUV's motion and sensor activation via an interface (typically a joystick or keyboard input) and real-time data feedback from the vehicle. Teleoperation has several advantages, including the following:

DOI: 10.1201/9781003737476-9

- **Immediate Response**: Operators can instantly adjust the vehicle's trajectory or sensor settings based on visual feedback.
- **Fine Control**: Direct control allows for precise manoeuvring in cluttered or dynamic environments.
- **High Situational Awareness**: Continuous input and monitoring enhance the operator's connection to the vehicle's current state.

However, teleoperation is labour-intensive and places a high cognitive load on the operator, particularly during long-duration missions. Additionally, the limitations of underwater communication—such as limited bandwidth and intermittent connectivity—can compromise the effectiveness of direct teleoperation.

9.1.1.2 Supervisory Control and Shared Autonomy

As AUV technology matured, control strategies shifted towards supervisory control and shared autonomy models. In these paradigms, the AUV executes pre-programmed mission plans autonomously, while the human operator oversees the overall mission and intervenes only when necessary. This approach offers several benefits:

- **Reduced Workload**: Operators can monitor multiple vehicles simultaneously without needing to control every movement.
- **Increased Scalability**: A single operator can supervise a fleet of AUVs, which is particularly important in large-scale operations.
- **Energy Efficiency**: Autonomous operation allows the AUV to optimise its energy use without frequent human intervention.

In supervisory control, the human–machine interface (HMI) typically provides a high-level summary of each vehicle's status, including key performance indicators such as battery levels, position, sensor health, and error states. Operators can then intervene—often through a graphical user interface (GUI) that allows them to issue commands (e.g., 'hold position' or 'return to base') if anomalies are detected.

9.1.1.3 Hybrid Interaction Models

Modern systems increasingly employ hybrid interaction models that blend direct control with supervisory oversight. These systems allow for different levels of control based on mission context. For instance, during critical phases of a mission (e.g., obstacle avoidance or emergency manoeuvres), the system may switch to a lower level of autonomy with direct operator input. Conversely, during routine survey operations, the system may rely on full autonomy with only periodic status updates from the vehicle.

Hybrid models require adaptable interfaces that can dynamically change the level of detail presented to the operator. For example, when the AUV is in safe cruising mode, the interface might display aggregated data (e.g., a heat map of the vehicle's path or a summary of sensor readings). However, if the system detects an anomaly, the interface could switch to a more detailed view, displaying live video feeds, sensor data streams, and diagnostic information.

9.1.2 Communication Channels

Operators interact with AUVs using a combination of communication channels:

- **Acoustic Communications**: Underwater, the primary mode of communication is acoustic. Acoustic modems can transmit telemetry data and command messages over distances of up to several kilometres, though at relatively low data rates. Acoustic communications are subject to environmental interference, multipath propagation, and latency.
- **RF and Satellite Links**: When AUVs surface or operate near a surface buoy, RF communications (including Wi-Fi or cellular networks) and satellite links may be used to transmit high-bandwidth data, such as video feeds or detailed sensor imagery.
- **Local Storage and Post-mission Download**: In many cases, AUVs store large amounts of data on board for later download. Operators then use post-mission analysis tools to review the data in greater detail.

In supervisory control scenarios, real-time data are often summarised and compressed to conserve bandwidth. For instance, rather than streaming full video feeds continuously, the system might transmit key frames or use event-driven reporting, where only anomalies or mission-critical events are reported in real time.

9.1.3 Monitoring Techniques

Effective monitoring of AUVs involves a combination of sensor data aggregation, data fusion, and visualisation techniques:

- **Dashboard Displays**: Mission control centres typically feature dashboards that consolidate critical parameters (e.g., battery life, depth, position, sensor status) into an easily digestible format. Graphical elements such as gauges, charts, and status indicators allow operators to quickly assess vehicle health.
- **Map-Based Interfaces**: Many systems use digital maps to display the AUV's real-time position, planned mission route, and environmental data. Interactive map interfaces enable operators to zoom in on areas of interest, view historical tracks, and identify potential hazards.
- **Alert Systems**: Automated alerts (visual, auditory, or haptic) notify operators when certain parameters exceed predefined thresholds. For example, an alert may be generated if the AUV's battery level drops below a critical value, or if the vehicle detects unexpected obstacles.
- **Data Fusion and Analytics**: Advanced systems integrate data from multiple sensors to produce higher-level interpretations of the underwater

environment. For example, combining sonar and video data can improve target detection and obstacle avoidance. Such fused data are then presented in a summarised form to the operator.

Together, these monitoring techniques allow human operators to maintain a high level of situational awareness even when controlling multiple AUVs, thus ensuring mission safety and success.

9.2 USER INTERFACES FOR MISSION CONTROL

Designing user interfaces for AUV mission control presents unique challenges due to the complex operational environment and the need to manage large amounts of heterogeneous data. This section outlines the key design principles and considerations that inform the development of effective mission control interfaces.

9.2.1 Design Principles

9.2.1.1 Situational Awareness

One of the most critical objectives of any mission control interface is to support operator situational awareness (SA). SA is defined as the perception of elements in the environment, comprehension of their meaning, and projection of their status in the near future. Effective interfaces should:

- **Display Critical Information Clearly**: Use well-organised dashboards and graphical displays that highlight essential parameters (e.g., vehicle position, energy levels, sensor health).
- **Hierarchical Data Presentation**: Information is organised hierarchically, with high-priority data at the top and options to drill down into detailed data. This reduces clutter and allows the operator to focus on mission-critical information.
- **Employ Multimodal Feedback**: Combine visual, auditory, and, when possible, haptic feedback to reinforce the operator's understanding of the AUV's state.
- **Minimise Cognitive Load**: Present information in a simplified and aggregated form, using methods such as heat maps or trend lines to reduce the need for constant interpretation of raw data.

Research on human–swarm interaction has shown that using visual abstractions like heat maps significantly reduces the operator's cognitive burden when managing multiple vehicles.

9.2.1.2 Usability and Ergonomics

The effectiveness of an interface is strongly linked to its usability. Reducing the cognitive workload of the human operator is a primary design goal. Research in human–swarm and human–robot interaction emphasises that user interfaces must be designed to support both high-level supervision and detailed control without overwhelming the operator. Future systems may incorporate artificial intelligence to predict operator workload and adjust interface complexity in real time. In mission-critical operations, the interface must be intuitive and easy to use, for which the salient features are listed below.

- **Consistency**: Interfaces should follow a consistent layout and design language to allow rapid familiarisation.
- **Responsiveness**: Feedback should be immediate, with minimal delay between operator command and system response.
- **Adaptability**: Interfaces may need to adapt dynamically depending on the phase of the mission. For example, an interface might provide more detailed controls during critical manoeuvres and revert to summary views during routine operation.
- **User Customisation**: Allowing operators to customise their displays (e.g., choosing which parameters to monitor, adjusting display scales) can greatly improve performance.
- **Error Prevention and Recovery**: The design should include mechanisms for error detection, clear notifications, and easy recovery options (e.g., one-click emergency stop or reset functions).

User studies in related domains (e.g., UAV swarm management) have shown that interfaces which allow a high degree of customisation and adapt to user preferences result in lower cognitive load and fewer operational errors.

9.2.1.3 Information Prioritisation

Given the volume of data generated by AUV sensors, prioritising the display of information is crucial. Not all data are equally important at every stage of a mission. Effective strategies include the following:

- **Layered Information Display**: Use hierarchical displays that show high-priority data at the top level and allow the operator to drill down for more detail as needed.
- **Alert and Notification Systems**: Highlight critical alerts using colour codes or flashing indicators to ensure the operator's attention is drawn to the most important issues.
- **Context-Aware Display**: The interface should dynamically adjust the information shown based on the current operational context. For example, during low-risk phases, the interface may display only summary statistics; during emergencies, detailed sensor readouts and control options should become available.

9.2.1.4 Transparency and Trust

Operators must trust the system to make autonomous decisions while still retaining ultimate control:

- **Data Transparency**: Display not only the decisions made by the autonomous systems but also the underlying sensor data and status information that led to these decisions.
- **Explainability**: When possible, provide brief explanations for automated actions so that operators can understand why certain commands were issued.
- **Intervention Options**: Ensure that operators have the ability to override automated decisions easily if they detect any discrepancies or anomalies.

Studies on human–machine interaction emphasise that interfaces that enhance transparency lead to increased trust and better performance in supervisory control scenarios (Soorati et al., 2021).

9.2.1.5 Safety and Reliability

Safety and reliability are paramount in AUV operations. Features towards these requirements are listed below.

- **Redundant Systems**: Many AUVs are designed with redundant sensors and communication channels to ensure continuous operation in the event of a component failure.
- **Error Detection and Recovery**: Automated error detection algorithms monitor the health of the system and trigger appropriate recovery procedures if anomalies are detected. In such cases, the interface notifies the operator and provides recommendations for corrective action.
- **Transparency**: The system should provide clear explanations for automated decisions to build operator trust. This includes displaying diagnostic information and the rationale behind specific actions taken by the AUV.
- **Interoperability Standards**: Adopting industry standards (e.g., JAUS, STANAG) enhances interoperability between systems, contributing to overall reliability and ensuring that operators can manage multiple systems with consistent interfaces.

9.2.2 Mission Control Interfaces

Technologies and tools used for mission control interfaces are discussed in this section.

9.2.2.1 Graphical User Interfaces (GUI)

Modern mission control centres use sophisticated GUIs to present real-time data, as listed below.

- **Dashboards**: Often include multiple panels such as status indicators, maps, video feeds, and sensor data graphs.
- **Touchscreen Interfaces**: Many systems incorporate touchscreen controls, enabling intuitive interactions such as drag-and-drop commands for setting waypoints.
- **Augmented Reality (AR) and Virtual Reality (VR)**: These technologies are increasingly used to present 3D visualisations of the underwater environment, providing operators with a more immersive view and improved depth perception. For instance, AR overlays can help highlight potential obstacles or mission-critical areas.

9.2.2.2 Software Platforms and Middleware

Several software platforms have been developed to support mission control:

- **QGroundControl**: An open-source ground control station originally designed for unmanned aerial vehicles (UAVs) but extended to support underwater vehicles. It integrates telemetry data, mission planning, and vehicle status information into one platform.
- **VideoRay Cockpit**: A proprietary interface used for remote operations of ROVs and AUVs that emphasises ease of use and robust data integration.
- **Custom Middleware**: In many cases, specialised middleware is developed to fuse data from multiple sensors, perform on-board processing, and relay information in an energy-efficient manner. This middleware must be designed with interoperability and scalability in mind.

9.2.2.3 Human Factors and Ergonomic Considerations

Ergonomic design is critical, and certain salient considerations are given here.

- **Display Layouts**: Well-organised and minimally cluttered layouts help reduce visual fatigue and enhance operator performance.
- **Font and Colour Choices**: High contrast, legible fonts, and carefully chosen colour schemes (e.g., red for alerts, green for normal status) ensure that operators can quickly interpret information even in high-stress situations.
- **Input Devices**: Operators often use specialised controllers or keyboards optimised for rapid command entry and error-free operation (Abdullah et al., 2024).

9.2.3 Case Studies and Examples

Several operational systems demonstrate effective mission control interfaces:

- The Mission Control System for the MARIUS AUV provides a good example of a Petri net-based framework that allows for both high-level mission planning and low-level control. The system enables operators to monitor vehicle status and intervene when necessary, using a combination of graphical displays and textual command interfaces (Oliveira et al., 1998).

FIGURE 9.1 Operator interfacing with UUV.

Source: **NIO Goa.**

- Recent innovations in AUV interfaces, such as those implemented on Kongsberg's HUGIN Endurance, highlight the shift towards adaptive interfaces that can switch between summary and detailed views based on the mission phase.

In summary, effective user interfaces for AUV mission control must support situational awareness, minimise operator workload, prioritise critical data, and offer transparency and intervention capabilities—all of which contribute to increased trust and better overall mission performance (Figure 9.1)

9.3 INTEGRATION AND SYSTEM-LEVEL CONSIDERATIONS

The integration of human interface design, energy management, and data processing forms the backbone of effective AUV mission control systems. In this section, we outline how these components are combined into a cohesive architecture that supports mission-critical operations.

9.3.1 Mission Control System Architecture

Modern mission control systems for AUVs typically include the following:

- **Operator Workstations:** These are equipped with adaptive GUIs that present real-time telemetry, sensor data, energy status, and diagnostic alerts. The design of these workstations is informed by human factors research to ensure minimal cognitive load.

- **On-Board Control Systems:** These include processors that perform sensor fusion, execute energy management algorithms, and implement adaptive communication protocols. On-board systems also handle local decision-making for tasks such as obstacle avoidance and mission adaptation.
- **Communication Middleware:** Middleware ensures seamless data exchange between the AUV and the operator. It supports multiple communication modes, automatically switching between low-power acoustic and high-bandwidth RF links as required.
- **Data Management Backends:** These systems handle both on-board storage and post-mission data transfer to cloud or centralised repositories. They support advanced analytics and long-term data mining.

Robust system architecture must integrate these components seamlessly. For instance, if the AUV's energy management system detects low battery levels, it can trigger an adaptive communication mode that reduces data transmission frequency. Simultaneously, the operator workstation is updated to reflect the energy status, enabling the operator to decide whether to alter the mission plan.

9.3.2 Key Design Considerations

Human interaction with AUVs is a multifaceted challenge that encompasses interface design, energy management, and data processing. Although AUVs have reached high levels of autonomy, human operators remain essential for mission planning, real-time monitoring, and emergency interventions. Effective mission control interfaces must be designed to support situational awareness, reduce cognitive load, and facilitate rapid decision-making while balancing the constraints imposed by limited energy resources. Key design considerations include the following:

- The adoption of adaptive, context-aware interfaces that dynamically change based on mission phase and operator workload.
- The implementation of intelligent communication protocols that conserve energy while ensuring that critical data are transmitted promptly.
- The use of on-board data fusion and edge computing techniques to summarise and compress voluminous sensor data, thus reducing the energy cost of data transmission.
- Integration with cloud-based analytics platforms to support post-mission analysis and long-term data management.

Future research will focus on further enhancing autonomy, improving the transparency of automated decision-making processes, and developing scalable interfaces for managing large swarms of AUVs. In parallel, advancements in battery technology, low-power processing, and standardised data protocols will help mitigate the energy constraints inherent in underwater operations.

9.4 DATA MANAGEMENT

The vast volumes of data generated during underwater surveys require robust data management systems. This section discusses how data are acquired, processed, transmitted, stored, and analysed. Data management solutions must address the challenges of on-board storage limitations, real-time processing constraints, and the need for robust post-mission data analytics.

9.4.1 Data Quality and Usefulness

The usefulness of underwater survey data depends on factors such as:

- **Resolution and Accuracy**: High-resolution data allow for detailed mapping and precise analysis, but may require more energy and bandwidth to acquire and transmit.
- **Coverage Area**: Extensive coverage is essential for large-scale assessments, though there is often a trade-off between area coverage and resolution.
- **Data Fusion**: Combining data from multiple sensors (e.g., sonar, video, and environmental sensors) enhances the overall quality and interpretability of the survey data.
- **Timeliness**: Real-time or near-real-time data is critical in dynamic environments or during emergency response scenarios.

Overall, high-quality survey data enable better decision-making in both scientific research and industrial applications, improving our understanding of the underwater environment and guiding effective resource management and infrastructure development.

9.4.2 Data Acquisition and On-Board Storage

AUVs are equipped with an array of sensors, and the data generated can be extremely large. Typical data management solutions at the vehicle level are described below.

9.4.2.1 Sensor Data Characteristics

AUV sensors typically generate data with the following characteristics:

- **High Resolution and Volume**: High-frequency sensors (e.g., SAS, MBES) can produce gigabytes of data per mission.
- **Heterogeneous Data**: Data types include sonar imagery, video feeds, environmental measurements, and navigational telemetry.
- **Real-Time Requirements**: Certain data (e.g., navigation and system health) must be processed and transmitted in real time, while other data may be stored for later analysis.

9.4.2.2 On-Board Storage Solutions

AUVs incorporate on-board storage solutions to temporarily hold raw data until it can be offloaded:

- **Solid-State Drives (SSD)**: Ruggedised SSDs are commonly used due to their high reliability and resistance to vibration and pressure.
- **Redundancy and RAID Configurations**: Multiple storage units are often configured in RAID to ensure data integrity even in the case of hardware failure.
- **Compression Algorithms**: Data are often compressed on board using both 'lossless' and 'lossy' techniques to maximise storage efficiency. Compression reduces the amount of data that must be transmitted, conserving both energy and bandwidth.

The integration of these storage systems is critical for ensuring that no data are lost during long missions, especially when real-time communication is not possible.

9.4.3 On-Board Processing and Edge Computing

Due to the constraints of underwater communication, AUVs need to perform significant on-board processing.

9.4.3.1 Edge Computing (Real-Time Data Processing)

On-board processing capabilities enable AUVs to analyse and summarise data in real time. Real-time data processing mentioned above is a pre-processing step that reduces energy consumption and allows for near real-time reporting of mission-critical data. Several techniques for 'data summarisation' help condense voluminous sensor data.

Deploying edge computing solutions allows complex computations to be performed on board, reducing the volume of data that needs to be transmitted. This is also referred to as 'local decision-making'. For example, machine learning algorithms can detect anomalies in sonar readings or classify objects in video feeds.

Some of the ways for data summarisation through real-time data processing are mentioned below. These summarisation techniques are vital for both real-time reporting and long-term storage. They ensure that operators receive the most relevant information without being overwhelmed by raw data.

- **Data Fusion**: Data from various sensors are fused to produce a coherent representation of the underwater environment. For example, sonar and video data may be combined to enhance object detection. The AUV can generate higher-level abstractions (e.g., environmental maps) that require less bandwidth to transmit.

- **Adaptive Sampling**: Algorithms (on-board) can adjust the sampling rate of sensors based on environmental conditions. When the environment is stable, the sampling rate may be reduced to conserve energy.
- **Feature Extraction**: Algorithms can extract key features (e.g., edges, contours, statistical summaries) from high-resolution images and sonar data.
- **Anomaly Detection**: By focusing on anomalous data—such as sudden changes in sonar echoes or temperature spikes—the AUV can prioritise critical data for immediate reporting.
- **Dimensionality Reduction**: Techniques such as principal component analysis (PCA) or auto-encoders can reduce the dimensionality of the data, making it easier to store and transmit.

9.4.4 Data Transmission Strategies

Due to the limited bandwidth of underwater communication channels, data transmission strategies are critical. These comprise communication protocols and 'energy-aware' transmissions.

9.4.4.1 Communication Protocols

- **Acoustic Modems**: Underwater acoustic modems are the primary means of communication, though they offer low data rates. Protocols must be robust against noise and multipath effects.
- **Store-and-Forward**: In situations where real-time data transmission is not feasible, AUVs may store data and offload it once a high-bandwidth link (e.g., when surfaced) is available.
- **Hybrid Communication**: When AUVs surface or communicate via tethered buoys, higher-bandwidth radio frequency (RF) or satellite communications can be used to transmit detailed data sets.

9.4.4.2 Energy-Aware Transmission

- **Dynamic Modulation**: Adaptive modulation schemes that adjust transmission parameters based on current energy levels and channel conditions help optimise power usage.
- **Event-Triggered Reporting**: Instead of continuous data streaming, systems often use event-triggered protocols where only critical data (or summaries) are transmitted in real time.
- **Scheduled Data Bursts**: Data may be stored on board and transmitted in bursts during predefined time windows when energy and communication conditions are optimal.
- **Prioritisation of Critical Data**: Data are prioritised based on importance—critical alerts and diagnostics are transmitted immediately, while less urgent data are buffered.

9.4.5 Post-mission Data Management

After the mission, large datasets must be offloaded and processed. Several options are available depending upon the volume, nature and time-criticality of the data.

9.4.5.1 Centralised or Distributed Storage

Data can be uploaded to central servers or cloud platforms where robust backup, search, and retrieval systems ensure data integrity.

In other cases, data management may involve distributed storage across multiple geographic locations, which is particularly useful for multinational projects or long-term environmental monitoring.

9.4.5.2 Big Data Analytics

The enormous volume of data generated by AUVs calls for advanced analytical techniques.

- **Big Data Frameworks/Cloud-Based Processing**: Once data are offloaded, cloud platforms provide the computational power required to analyse and interpret large datasets. Frameworks such as Hadoop and Spark are used for data mining and trend analysis.
- **Machine Learning**: Advanced algorithms are applied to the data to detect patterns, classify features, and predict future environmental changes. This analysis is critical for refining future survey missions and improving the design of AUVs.
- **Visualisation Tools**: Data visualisation software transforms complex datasets into intuitive graphics—such as 3D maps, heat maps, and temporal trend charts—allowing scientists and engineers to extract actionable insights.

9.4.6 Data Fusion and Integration

Data fusion techniques integrate data from multiple sensors to provide a comprehensive picture of the underwater environment. Effective data fusion can reduce redundancy, improve accuracy, and facilitate better decision-making.

By combining overlapping datasets, the system can eliminate redundancy and focus on the most relevant information. Fusing data from different sensors (e.g., sonar and video) often leads to more accurate and robust measurements. A coherent, fused dataset helps operators make informed decisions during mission execution and in post-mission analyses.

Integrating data over time allows for trend analysis and improves the accuracy of environmental models. This is particularly useful in dynamic environments where conditions change rapidly.

In summary, robust data management solutions in AUV operations encompass on-board processing and summarisation, adaptive transmission protocols, centralised and distributed data storage architectures, and advanced big data analytics. Together, these solutions enable efficient handling of large data volumes while supporting real-time operational requirements and post-mission analysis.

9.5 COLLABORATIVE AUV OPERATIONS

9.5.1 Necessity

In many underwater missions, such as seabed mapping, environmental monitoring, or surveillance, a single AUV may take many hours or days to complete its task. On the other hand, a coordinated swarm could divide the area of interest into sectors, each covered by an individual agent. This division of labour not only reduces mission duration but also increases spatial resolution and data redundancy.

The major motivations for collaborative AUV operations can be summarised as follows, which are then discussed further.

- Enhanced Coverage and Efficiency.
- Robustness and Redundancy.
- Adaptability and Real-Time Decision-Making.
- Scalability and Cost-Effectiveness.

One of the primary drivers behind the shift towards swarm operations is the ability to cover larger areas in a shorter time span. For instance, by using multiple agents that can share data in real time, it becomes possible to detect and localise anomalies (e.g., pollutant plumes or underwater hazards) more accurately than would be feasible with a solitary unit. In addition, coordinated path planning algorithms allow the swarm to adapt dynamically to new information, redistributing agents as needed to areas of interest.

Swarm operations inherently offer greater resilience compared to single-vehicle systems. In a well-designed swarm, the failure or temporary malfunction of one or a few agents does not compromise the overall mission. Instead, the remaining vehicles can reconfigure their trajectories and roles to compensate for the loss. This redundancy is especially critical in harsh underwater environments where reliability is paramount, and recovery operations may be expensive or impractical.

The distributed nature of swarm control allows individual agents to make decisions based on local sensor data, while still contributing to a global mission objective. Advances in machine learning and decentralised control algorithms have enabled AUVs to perform tasks such as obstacle avoidance, target tracking, and dynamic re-tasking without constant human supervision. This level of autonomy is key to adapting to rapidly changing underwater conditions (such as variable currents, obstacles, or communication disruptions) and ensures that the swarm can operate effectively in complex and unstructured environments (Hambling, 2016).

Swarm operations are typically based on modular designs where relatively low-cost vehicles work in concert. This modularity makes it economically feasible to deploy large numbers of AUVs. In contrast to high-end systems that are expensive to maintain, a swarm composed of inexpensive, mass-produced vehicles can achieve significant coverage and redundancy. Moreover, advances in miniaturised sensors and energy-efficient designs further contribute to the scalability of these systems.

Table 9.1 summarises key experimental projects in the area of collaborative AUV operations, highlighting quantitative parameters demonstrated in each.

TABLE 9.1
Examples of Collaborative AUV Projects (Swarm Operations) Demonstrated

Project Name	Agents	Operations Area/Distance	Endurance	Communication Method	Notable Characteristics
SWARMs (2016–2019)	8	Up to 10 km^2 area; 1–2 km range (acoustic)	~24 hours	Hybrid (acoustic, optical)	Real-time data fusion, decentralised control, adaptive mission planning
AQUA Swarm (2017–2019)	5–6	3–5 km coastal operations	~20 hours	Primarily acoustic	Energy-efficient operations, real-time adaptive mission management
MARUS (2015–2018)	4–5	4 km operational range	~18 hours	Acoustic modems (multi-hop)	Formation control, cooperative obstacle avoidance, robust navigation
EuRo-AUV (2018–2020)	6	~6 km operational distance	~22 hours	Acoustic + satellite relay	Multi-mission adaptability, integrated sensor networks
Cooperative Ocean Survey (COOS) (2019–2021)	7	5–7 km area coverage	~20–24 hours	Acoustic and RF (near-surface)	High-resolution mapping, adaptive reconfiguration, redundant sensing

9.6 ROV OPERATIONS

Remotely Operated Vehicle (ROV) systems consist of multiple tightly integrated elements, each essential to effective subsea performance. At the core is the vehicle itself, which incorporates propulsion units, on-board power distribution, navigation and sensing systems, tool and manipulator interfaces, and a pressure-rated structural frame. The vehicle is connected to the support vessel via an umbilical, which supplies electrical power—typically in the range of several hundred kilowatts for heavy-duty systems—and enables two-way transmission of control commands, telemetry data, and real-time video.

Shipboard infrastructure complements the subsea system and normally includes pilot control stations, power generation and conditioning equipment, umbilical handling and storage systems, and dedicated launch and recovery arrangements designed to safely deploy and retrieve the vehicle (Christ, 2011).

9.6.1 Mission Deployment

ROV capability is now common on a range of ocean-going ships, including cable-laying ships, hydrographic survey vessels, offshore support ships, and scientific research platforms. Many of these vessels are configured to operate more than one class of ROV, allowing flexibility in mission planning and execution.

The selection of ROVs versus alternative intervention methods is influenced by several parameters, such as operating depth, seabed characteristics, required burial or intervention depth, prevailing environmental conditions, project schedule constraints, and the capabilities of the available vessel. In near-shore or shallow water environments, diver-based operations or specialised mechanical systems may sometimes be feasible. However, at depths exceeding approximately 100 m, UUVs become the only practical means of inspection, maintenance, and repair for industrial, scientific, or defence-related seabed infrastructure.

Operational depth and endurance are constrained by the umbilical connecting the ROV to the mother ship. As the umbilical length increases, electrical losses reduce the effective power available at the vehicle, limiting propulsion output and tool performance (CURV III, 2016). Standard ROV systems typically employ umbilicals of around 2,000 to 3,000 m, which restricts their usable depth. Extending operational range necessitates heavier and thicker umbilicals, which introduce handling difficulties, increased system weight, and potential impacts on vessel stability (MapYourTech, 2025).

9.6.2 Mission Duration

The length of individual ROV missions is governed by a combination of human, mechanical, and operational factors. Although the vehicle itself can function continuously as long as power is available, sustained operations are limited by pilot fatigue, progressive wear of mechanical components, and thermal buildup in hydraulic and electrical systems. As a result, ROV deployments are commonly planned for periods of approximately 8 to 16 hours, followed by recovery for inspection, servicing, and preparation for subsequent dives. For longer campaigns, multiple pilot teams may be rotated to maintain safe and effective operations.

Environmental conditions further constrain mission planning. While the ROV may remain stable and functional underwater during adverse weather, safe launch and recovery require acceptable sea states. In regions with narrow weather windows, project schedules must account for potential downtime and vessel standby periods. For time-critical subsea activities, the financial consequences of weather-related delays can be substantial, often exceeding the direct cost of ROV operations themselves.

9.6.3 Coordination and Safety

Close coordination between ROV activities and surface vessel operations is essential. The umbilical presents both a physical and operational hazard, with the potential to foul propulsion systems, interfere with cable handling equipment, or restrict vessel manoeuvrability. Although it need not maintain its position dynamically, the mother ship is usually constrained in its movements as long as its ROV is deployed (Christ, 2011).

Safety considerations underpin all aspects of ROV deployment. The umbilical constitutes a critical single-point vulnerability; damage or severance can result in loss of the vehicle on the seabed. While rigorous inspection and handling procedures significantly reduce this risk, it cannot be entirely eliminated. High-power electrical transmission through the umbilical also introduces hazards to personnel and adjacent systems. Accordingly, operational procedures mandate strict electrical isolation, lockout protocols, and coordination with shipboard activities during ROV deployment.

9.6.4 ROV Crew

Human expertise remains a decisive factor in ROV performance. The skill and experience of pilots directly influence productivity, equipment longevity, and overall mission success. Seasoned operators develop a refined understanding of vehicle dynamics, anticipate abnormal conditions, and adapt control strategies to maximise efficiency and safety (Griffiths, 2003).

In addition to pilots, supervisors, technicians, and deck crews play critical roles and require specialised training. Umbilical handling, in particular, demands disciplined procedures, as improper management can lead to equipment damage or hazardous situations. Training programmes emphasise electrical safety, emergency response, and fault-recovery procedures.

Launch and recovery operations require precise coordination between deck personnel and control room staff to minimise risks to both equipment and personnel.

Advances in automation have begun to reduce operator workload and improve consistency. Contemporary ROV systems commonly feature functions such as automatic depth and heading control, sensor-based cable tracking, and automated burial depth regulation. Looking ahead, increased autonomy may enable full burial operations to be conducted under supervisory control. Nevertheless, tasks involving repair, manipulation, and fault diagnosis are likely to continue relying on human judgement due to their complexity and unpredictability.

9.6.5 Maintenance

Since offshore ROV operations are often conducted in remote locations and support mission-critical activities such as cable repairs or offshore platform repairs, reliable operations are essential. A comprehensive inventory of spare parts is therefore indispensable. Typical spares include hydraulic components, electrical connectors and penetrators, thruster assemblies, cameras and lighting units, navigation sensors, and mission-specific tooling. The spare strategy must strike a balance between operational resilience, available storage space, and overall mobilisation cost.

Routine maintenance between deployments is essential to detect degradation before it leads to failure. Standard inspection procedures include checks for seal integrity and water ingress, verification of thruster and electrical performance, assessment of hydraulic system parameters, confirmation of imaging and lighting functionality, and calibration of navigation and sensor systems.

Post-dive servicing periods allow for component replacement, system testing, and readiness verification. Detailed maintenance records support trend analysis and help identify recurring faults linked to design limitations or procedural shortcomings.

9.6.6 Good Practices

Effective ROV operations are built on aligning vehicle capabilities with mission demands, enforcing disciplined deployment and recovery procedures, investing in personnel training and experience, accommodating environmental and weather constraints, coordinating closely with wider vessel operations, and sustaining robust maintenance and spare programmes.

Despite ongoing technological advances, the core foundations of ROV operations remain unchanged. Reliable outcomes are achieved when skilled teams operate well-maintained systems in accordance with established procedures, regardless of the specific mission or operating environment (MapYourTech, 2025).

9.7 UNDERWATER NETWORKS

9.7.1 Internet of Underwater Things (IoUT)

The future of collaborative undersea operations would involve networked sensor and vehicle systems, referred to as Internet of Underwater Things (IoUT) or Subsea Internet of Things (SIoT) (Nkenyereye, Nkenyereye, & Ndibanje, 2024). Such underwater networks would need to rely largely on acoustic communication (Figure 9.2). In the earlier chapters, we mentioned the relevance of IoUT for optimising AUV energy consumption and the significance of underwater communication for realising this concept. Initial efforts in this area were by the US Office of Naval Research

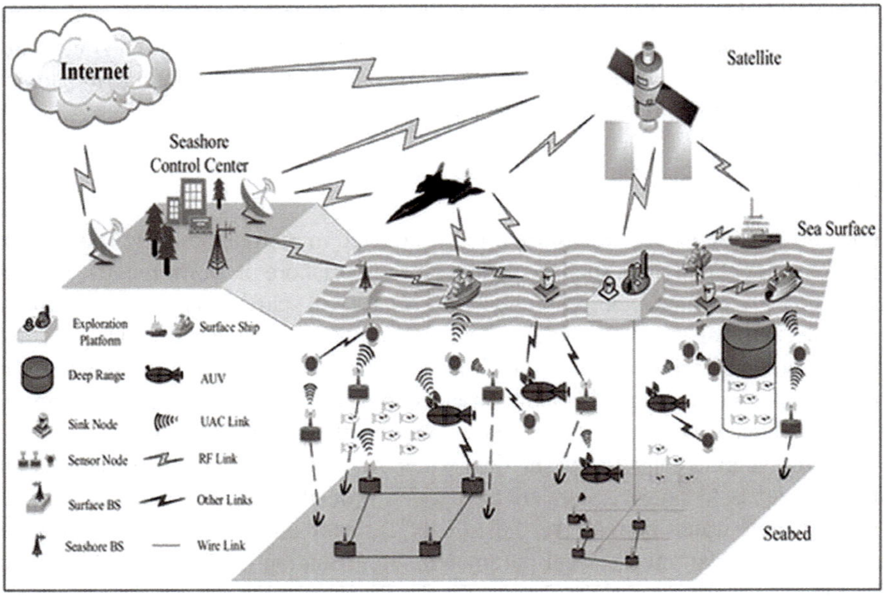

FIGURE 9.2 Model of IoUT with various types of communication networks.

Notes: **Qiu, Zhao, Zhang, Chen, and Chen (2020).**

(ONR) and DARPA through the PLUSNet (Persistent Littoral Undersea Network) project that started in 2005 (Stewart & Pavlos, 2006).

9.7.2 PLUSNet

PLUSNet was an applied research programme funded by the Office of Naval Research from 2005, integrating more than a dozen demonstrated technologies to test and evaluate the merits of a prototypical, persistent, automated, tactically and environmentally adaptive sensor network (Figure 9.3).

This network of fixed and mobile sensors and underwater communications nodes was to employ advanced low-power signal processors for contact detection, classification, and localisation (Figure 9.4) (Stewart & Palvos, 2006). It would exercise different levels of autonomy from a remote central controller, including contact reporting, collaborative tracking, and contact hand-off to an adjacent sensor network. It was envisaged to demonstrate the coupling of an SSGN with this off-board sensor network (consisting of mobile AUVs as well as fixed sensors) to develop a persistent anti-submarine warfare capability against modern diesel submarines in shallow waters (Stewart & Pavlos, 2006).

FIGURE 9.3 Concept of persistent littoral undersea network (PLUSNet) of ONR and DARPA.

Notes: **Martin, 2005; https://nps.edu/documents/103424733/107333295/Autonomous+Platfors+in+Persistent+Littorl.pdf**

9.7.3 SeaWeb

In 2010, researchers at the US Naval Postgraduate School in Monterey, California, unveiled Seaweb, a system similar in concept to PLUSNet, after more than ten years of development. Seaweb is an acoustic communications network which uses through-water acoustic modems to connect up a large number of sensor nodes to a surface gateway node, allowing real-time digital communication between data gathering sensors and land-based command centres (Lo, 2012).

Their original goal was to create a network of distributed sensors for detecting quiet submerged submarines in littoral waters where traditional ASW surveillance is challenged by complex sound propagation and high noise (Vandenberg, 2010).

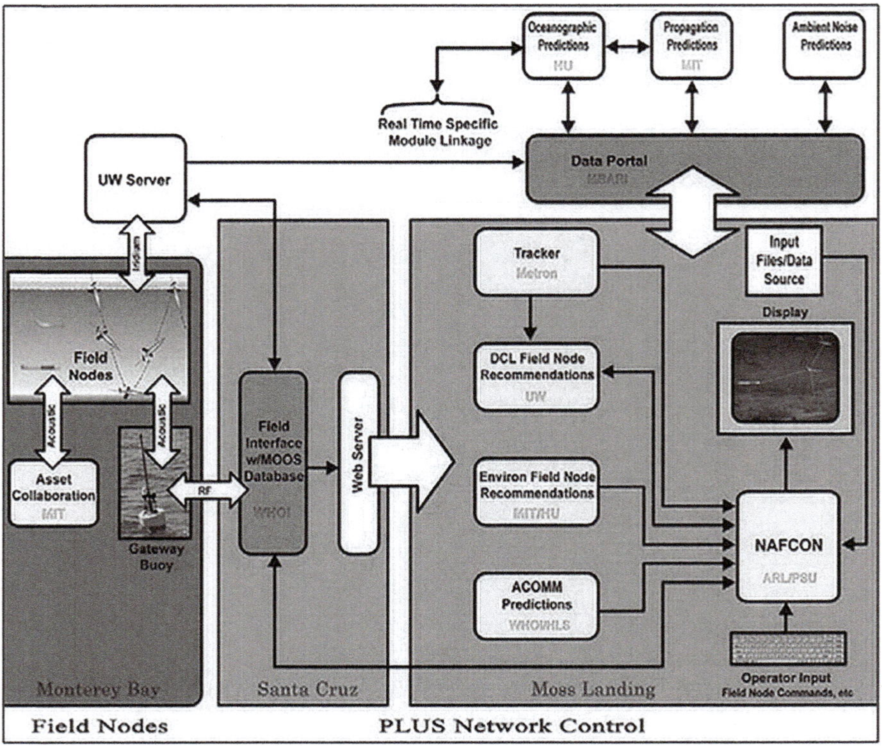

FIGURE 9.4 PLUSNet shore component configuration and connections.

Notes: **Stewart and Pavlos (2006); http://mseas.mit.edu/archive/PLUSNet/5stewamsPAF2.pdf**

In this concept, communication between non-tethered UUVs and submarines is provided via underwater digital acoustic communications (ACOMMS). ACOMMS is a short-range, modem-like communication method that is often left unencrypted (though it could easily be encrypted). In 2005, Seaweb ran three experiments using their expendable node technology to aid in the navigation of several types of UUVs. These tests showed that undersea ACOMMS technologies using multiple scattered nodes are a viable solution for naval communication. The next step for Seaweb was to utilise the sensor nodes to communicate to and from UUVs and submarines, as well as relay mission-specific information to and from satellites via various surface assets, as shown in Figure 9.5 (Vandenberg, 2010).

In an article entitled ‘Enabling Undersea ForceNET with Seaweb Acoustic Networks’ in 2003, author Joseph Rice of SPAWAR San Diego forecasted that:

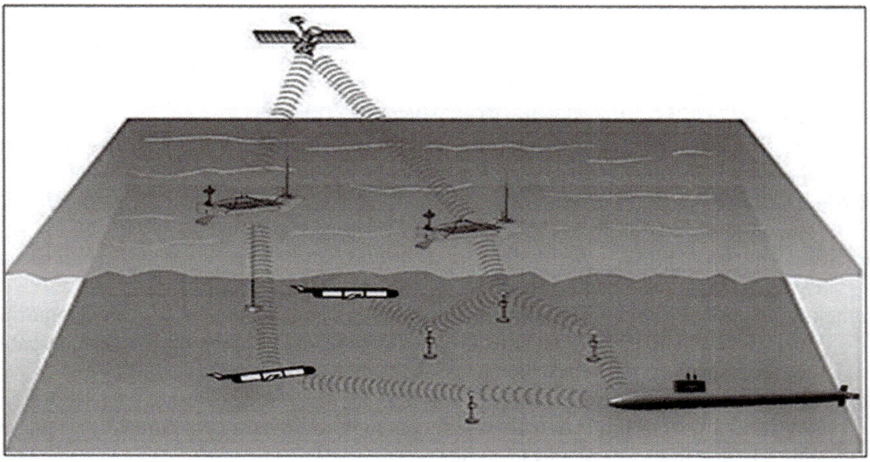

FIGURE 9.5 Representation of 'SeaWeb' acoustic communication and navigation network.

Notes: **Rice and Green (2008); https://www.researchgate.net/publication/4373267_Underwater_Acoustic_ Communications_and_Networks_for_the_US_Navy's_Seaweb_Program**

> Undersea, off-board, autonomous systems will enhance the war-fighting effectiveness of submarines, maritime patrol aircraft, amphibious forces, battle groups, and space satellites. Wide-area sensor grids, leave-behind multi-static sonar sources, mine-hunting robots, and AUVs are just a few of the battery-powered, deployable devices that will augment space and naval platforms (Vandenberg, 2010).

These are visionary but ambitious goals of networking not only underwater nodes but linking them to assets on the surface, in the air, and in space (Nichols et al., 2020) If realised, these could truly be 'force multipliers' in the maritime domain (Ray, Singh, & Seshadri, 2011).

9.8 CONCLUSION

In conclusion, the successful integration of human–machine interfaces, energy management strategies, and robust data processing systems is vital for the next generation of AUV missions. As the technology continues to evolve, the combined efforts of researchers and industry practitioners will lead to more efficient, reliable, and user-friendly systems that can operate effectively in the challenging underwater environment. The continuous feedback loop between operational data, human intervention, and automated system responses will ultimately enhance both mission safety and overall performance.

REFERENCES

Abdullah, A., Chen, R., Blow, D., & Uthai, T. (2024, December). Human-Machine Interfaces for Subsea Telerobotics: From Soda-straw to Natural Language Interactions. *eprint arXiv:2412.01753*. https://doi.org/10.48550/arXiv.2412.01753

Christ, R. S. (2011). *The ROV Manual: A User Guide for Observation Class Remotely Operated Vehicles*. Netherlands: Elsevier Science.

CURV III. (2016, April 21). United States Naval Undersea Museum. https://navalundersea-museum.org/curv-iii/

Griffiths, G. (Ed.). (2003). *Technology and Applications of Autonomous Underwater Vehicles*. Taylor & Francis. https://doi.org/10.1201/9780203023249

Hambling, D. (2016, March). The Inescapable Net: Unmanned Systems in Anti-Submarine Warfare. *BasicInt*. https://basicint.org/publications/david-hambling/2016/inescapable-net-unmanned-systems-anti-submarine-warfare

Lo, C. (2012, April 30). *Persistent Littoral Surveillance: Automated Coast Guards*. Retrieved August 2024, from https://www.naval-technology.com/features/featurenavy-persistent-littoral-surveillance-auvs-uuvs/?cf-view

MapYourTech. (2025, December 7). ROV Operations in Subsea Cable Installation. *MapYourTech*. https://mapyourtech.com/rov-operations-in-subsea-cable-installation/

Martin, D. L. (2005, October 6). Autonomous Platforms in Persistent Littoral Undersea Surveillance: Scientific and Systems Engineering Challenges. *Naval Postgraduate School*. https://nps.edu/documents/103424733/107333295/Autonomous+Platfors+in+Persistent+Littorl.pdf

Moore, S., Bohm, H., & Jensen, V. (2010). *Underwater Robotics: Science, Design and Fabrication*. Monterey, CA: Marine Advanced Technology Education Center.

Nichols, R. K., Mumm, H. C., Lonstein, W. D., Ryan, J. J. C. H., Carter, C. M., Hood, J.-P., Shay, J. S., Mai, R. W., & Jackson, M. J. (2020). *Unmanned Vehicle Systems & Operations on Air, Sea, Land*. New Prairie Press. https://doi.org/10.4148/2334-0402.1030

Nkenyereye, L., Nkenyereye, L., & Ndibanje, B. (2024). Internet of Underwater Things: A Survey on Simulation Tools and 5G-Based Underwater Networks. *Electronics (Special Issue: Artificial Intelligence Empowered Internet of Things), 13*(3). https://doi.org/10.3390/electronics13030474

Oliveira, P., Pascoal, A., Silva, V., & Silvestre, C. (1998). Mission Control of the MARIUS AUV: System Design, Implementation, and Sea Trials. *International Journal of Systems Science, 29*(10), 1065–1080. https://doi.org/10.1080/00207729808929598

Qiu, T., Zhao, Z., Zhang, T., Chen, C., & Chen, C. (2020). Underwater Internet of Things in Smart Ocean: System Architecture and Open Issues. *IEEE Transactions on Industrial Informatics, 16*, 4297–4307. https://ieeexplore.ieee.org/abstract/document/8863935

Ray, A., Singh, S., & Seshadri, V. (2011). Underwater Gliders: Force Multipliers for Naval Applications. *Warship 2011: Naval Submarines and UUVs*. Bath, UK: Royal Institution of Naval Architects. https://doi.org/10.3940/rina.ws.2011.03

Rice, J. A., & Green, D. (2008). *Underwater Acoustic Communications and Networks for the US Navy's Seaweb Program*. Sensor Technologies and Applications SENSORCOMM '08. San Diego. https://doi.org/10.1109/SENSORCOMM.2008.137

Soorati, M., Clark, J., Ghofrani, J., Tarapore, D., & Ramchurn, S. (2021). Designing a User-Centered Interaction Interface for Human–Swarm Teaming. *Drones, 5*(4), 131. https://doi.org/10.3390/drones5040131

Stewart, M. S., & Pavlos, J. (2006). A Means to Networked Persistent Undersea Surveillance. *Submarine Technology Symposium*. Retrieved September 2024, from http://mseas.mit.edu/archive/PLUSNet/5stewamsPAF2.pdf

Vandenberg, T. D. (2010). *Manning and Maintainability of a Submarine Unmanned Undersea Vehicle (UUV) Program: A Systems Engineering Case Study.* Monterey, CA: Naval Postgraduate School. Retrieved January 2025, from https://apps.dtic.mil/sti/tr/pdf/ADA531594.pdf

10 Deployment, Manipulators, and Actuators

In this chapter, we examine the means for deploying UUVs into the sea, which are termed as 'Launch and Recovery Systems' (LARS). In the case of manned submersibles, these are the evolutions that pose the maximum risk to personnel. In the case of unmanned systems, these pose the maximum risk for loss of the expensive and complex vehicle. The first part of the chapter covers various types of LARS, their design considerations, and the range of features.

The subject of robotics in general is often dominated by the design of manipulators, their control system, and their sensors and effectors. While manipulators are critical for most robotic missions on land, they form only one component of the myriad challenges in underwater robotics. They do become critical wherever intervention tasks are to be executed underwater, whether by ROVs or AUVs.

At the end of the chapter, soft actuators (including smart materials) and their application to soft underwater robotics are introduced.

10.1 LAUNCH AND RECOVERY SYSTEMS

Underwater Vehicle Launch and Recovery Systems (LARS) are engineered to safely deploy and retrieve UUVs from a support platform. These systems can be integrated into the hull or deck of a vessel, installed on dedicated platforms, or even be part of unmanned surface vehicle (USV) operations. LARS must ensure that vehicles are transferred through the splash zone safely, minimise the risk of damage or loss, and enable rapid turnaround for multiple missions.

10.1.1 Evolution of Systems

Early systems were simple cranes or davits used to lower vehicles by hand into the water (Ballard & Drew, 2021). Even now, man-portable AUVs (less than about 30 kg) can just be dropped over the side from a boat, but recovering them requires precise homing and is a bigger challenge.

LAR systems have evolved as underwater vehicles became more complex and capable of deeper dives (Ray, Singh, & Seshadri, 2011). In the 1960s and 1970s, as ROVs and AUVs began to be used for tasks such as cable surveys and underwater salvage, specialised systems were developed that featured mechanisms such as A-frames and telescopic arms (Rumson, 2021). Deployment of ROVs has challenges

DOI: 10.1201/9781003737476-10

FIGURE 10.1 Example of launch and recovery system for UUV.

Source: **WHOI.**
Notes: **WHOI.**

related to handling a long umbilical mounted on a winch, which needs to be paid out (and recovered) for hundreds (or even thousands) of metres (Pantheeradiyil, 2024; Christ, 2011).

Organisations such as the Woods Hole Oceanographic Institution (WHOI) and companies like NOV-BLM led the way by designing integrated systems that could operate in harsh sea conditions and support vehicles with increasing payload and performance requirements (Figure 10.1) (NOV, 2025; WHOI, 2025).

10.1.2 Process and Sequence

During launch, the UUV secures to the LARS (for example, an A-frame) while initial system checks verify that all subsystems function correctly. The vehicle then descends through the splash zone, where wave action can impose significant forces on both the UUV and the lift wire. Motion compensation systems actively adjust wire tension to minimise shock loads. Once clear of surface wave effects, the UUV transitions to powered descent using its own thrusters. In the case of an ROV, as it dives, its umbilical cable pays out from dedicated winches. Operators monitor descent rate, vehicle attitude, and umbilical management throughout the deployment sequence.

Recovery reverses this sequence but demands greater caution, especially for ROV in case of an increase in weight during bottom operations and potential seabed material accumulation on the vehicle. The ROV first swims clear of the seabed, then transitions to lift wire support for ascent through the water column. Surface crews monitor vehicle ascent rate, umbilical tension, and vessel motion to optimise recovery timing.

In case of AUVs, the vessel floats to the surface and can be immobile, in which case it would require a diver to enter the water for physically connecting it. Alternatively, it could be actively guided into a recovery arm where it can be secured in place for hauling on board.

In either case (ROV or AUV), the LARS motion compensation system remains active throughout recovery, reducing dynamic loads that could damage equipment or endanger personnel. Advanced LARS capabilities extend operational weather windows compared to crane-based launch and recovery, which requires nearly calm conditions for safe UUV handling.

10.2 CHALLENGES IN LAUNCH AND RECOVERY

LAR systems face a host of challenges due to the nature of the marine environment and the operational requirements of underwater vehicles. These challenges can be broadly categorised into environmental, operational, technical, and human factors.

10.2.1 Environmental Challenges

10.2.1.1 Sea State and Weather Conditions

The most significant challenges for any LARS are the unpredictable and often severe marine environment. High sea states characterised by large waves, strong currents, and turbulent conditions make it difficult to maintain precise control over the deployment and recovery process. For instance, a system designed for operations in sea state 2 may not be reliable in sea state 5, where wave heights and vessel motions can drastically alter the dynamics of the splash zone. Therefore, LARS must be robust enough to function in a range of environmental conditions while ensuring the safety of both the UUV and the support vessel (Figure 10.2).

10.2.1.2 Dynamic Vessel Motion

Support vessels are seldom stationary, particularly during operations at sea. The motion induced by waves, tides, and wind means that the launch and recovery system must accommodate significant heave, pitch, and roll (Wynn et al., 2014). A well-designed LARS will include heave compensation mechanisms—such as shock absorbers, telescopic arms, or dynamic winch systems—that can adapt to these movements in real time, ensuring that the vehicle remains stable during the transition through the water's surface (Sagar et al., 2025) (Figure 10.3).

10.2.1.3 Hydrodynamic Forces

The process of lowering or hoisting a vehicle through the water subjects the system to complex hydrodynamic forces. Drag, lift, and turbulence can cause unwanted

FIGURE 10.2 Israeli AUV Blue Whale being launched.

Source: **German MoD.**
Notes: **https://armyrecognition.com/news/navy-news/2025/greece-to-use-israeli-blue-whale-unmanned-submarine-to-improve-maritime-intelligence-gathering; Accessed May 25.**

FIGURE 10.3 LARS and AUV (Autosub6000) attached through motion compensation mechanism.

Notes: **Wynn et al., 2014; Sagar et al., 2025; https://ars.els-cdn.com/content/image/1-s2.0-S0025322714000747-gr1.jpg**

oscillations or misalignment, which in turn may damage sensitive instruments on the vehicle or the recovery system itself. Hydrodynamic design considerations are therefore critical; LARS designers use computational fluid dynamics (CFD) and physical scale model testing to optimise system shapes, materials, and motion-compensation features.

10.2.2 Operational Challenges

10.2.2.1 Safety Considerations

Safety is paramount during launch and recovery operations. A malfunction or misalignment during these operations can result in damage to expensive vehicles, injury to personnel, or even catastrophic loss of life. Safety features in LARS often include fail-safe mechanisms, redundant systems, and real-time monitoring of load, tension, and position. Additionally, remote or autonomous systems are increasingly employed to reduce human exposure to hazardous conditions on deck during rough seas.

10.2.2.2 Load Management

Underwater vehicles vary widely in size and weight. The LARS must be capable of managing dynamic loads associated with heavy vehicles as well as lighter ones. This includes not only the static weight of the vehicle but also dynamic forces such as impact loads from wave action and inertial loads during acceleration and deceleration. The use of advanced winch systems, counterweights, and load-distribution mechanisms is common in modern LARS designs.

10.2.2.3 Integration with Vehicle and Vessel Systems

LARS do not operate in isolation; they must integrate seamlessly with both the underwater vehicle and the support vessel's systems. This involves compatibility with the vehicle's attachment points, communication protocols, and control systems. At the same time, the LARS must be integrated into the vessel's structure without compromising its stability or operational capabilities. Achieving this level of integration often requires customised engineering solutions tailored to the specific vehicle and vessel combination. The ability to fit LARS to smaller surface ships presents opportunities to operate UUVs from USVs (Figure 10.4).

10.2.3 Technical Challenges

10.2.3.1 Mechanical Complexity

Modern LARS often incorporate sophisticated mechanical components such as telescopic arms, winches, cranes, and shock-absorbing cradles. These components must be precisely engineered to withstand the harsh conditions of the marine environment while performing reliably over many cycles. Mechanical failures, such as cable fatigue, hydraulic leaks, or component wear, are significant concerns. As such, design approaches must include rigorous testing protocols and the use of high-grade, corrosion-resistant materials.

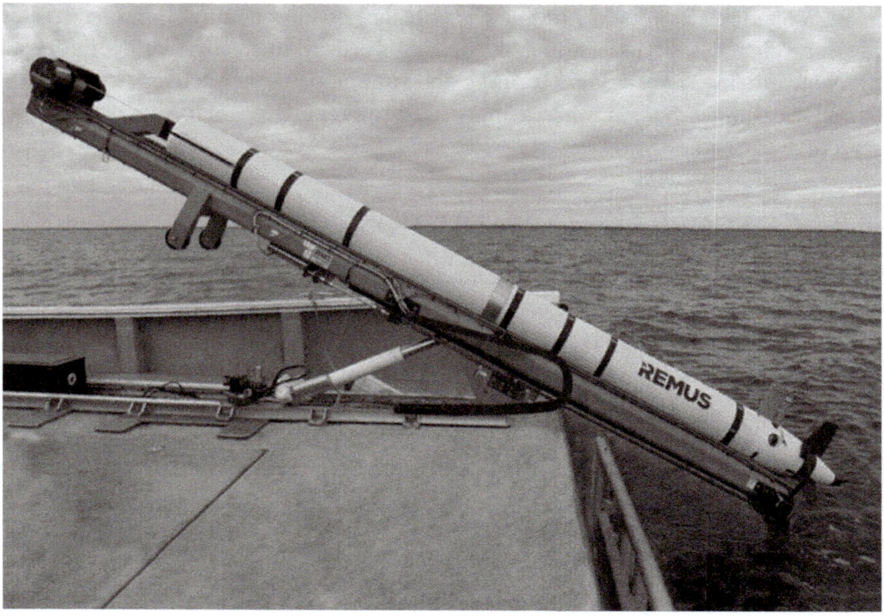

FIGURE 10.4 Sea Launcher LARS (by Hydroids) deploying a REMUS600 AUV.

Notes: **https://bluezonegroup.com.au/announcements/remus-autonomous-launch-recovery/**

10.2.3.2 Control and Automation

Automation is increasingly common in LARS to improve safety and efficiency. However, developing control algorithms that can manage the complex interactions between vessel motion, environmental forces, and mechanical system dynamics is challenging. Feedback from sensors such as accelerometers, gyroscopes, GPS, and acoustic positioning systems must be integrated into a coherent control system. This system needs to adjust in real time to disturbances and unexpected events during launch and recovery.

10.2.3.3 Communication and Data Transfer

Effective communication between the LARS, the support vessel, and the underwater vehicle is critical, especially when systems are remotely operated. Underwater communications, often relying on acoustic modems, have limited bandwidth and can be affected by environmental noise. Ensuring reliable data transfer for control signals and video feedback requires robust communication systems and often the use of redundant channels.

10.2.4 Human Factors and Operational Procedures

The interface between human operators and automated systems is another critical challenge. Even with high levels of automation, human oversight remains essential during complex recovery operations. The design of the operator control interface, including clear visual and auditory feedback, is crucial to ensure that operators can intervene quickly if necessary. Training and simulation tools, such as virtual reality and physical mock-ups, are increasingly used to prepare crews for handling LARS in a range of conditions.

10.3 CATEGORIES OF LARS SOLUTIONS

LARS solutions vary widely in design, reflecting differences in application, vehicle size, operational environment, and platform type. Broadly, these solutions can be categorised into four main types.

10.3.1 Fixed Platform Systems

Fixed platform systems are typically integrated into the deck or hull of a vessel. They include the following:

- **Crane-Based Systems**: These systems use shipboard cranes to lift the underwater vehicle directly from the water. They are common on larger vessels where deck space is not a limiting factor. The simplicity of a crane makes it a robust solution, although it may require additional mechanisms (e.g., shock absorbers or specialised slings) to manage dynamic loads (Figure 10.5).
- **A-Frame Systems**: An A-frame mounted on the stern or side of a vessel provides a guided structure for lowering and raising vehicles. The A-frame can help align the vehicle and protect it during the transition through the splash zone. Many systems incorporate heave compensation features to mitigate vessel motion (Figure 10.6).

FIGURE 10.5 Crane with padded hat and release/capture device to minimise free swing angle, which maintains AUV attitude.

Source: **MBARI.**
Notes: **Polar AUV Guide: Deployment and Recovery; https://astrobiology.nasa.gov/news/diving-under-icy-waters/**

FIGURE 10.6 LARS A-frame arrangement for deploying submersibles.

Source: Royal IHC.
Notes: **https://www.royalihc.com/defence/defence-equipment/launch-and-recovery-systems/submarine-rescue-vehicle-launch-and-recovery**

- **Telescopic Arms and Gantries**: These systems offer a compact solution that can extend over the water and retract once the vehicle is in the water. Telescopic designs are particularly useful in situations where deck space is limited (Figure 10.7).

10.3.2 Mobile Vessel Systems

Mobile vessel systems are those integrated into support vessels that may not be dedicated solely to underwater vehicle operations:

- **Dedicated Launch and Recovery Vessels (LRVs)**: These are ships built specifically or heavily modified to support underwater operations. They often include integrated cranes, winches, and dynamic positioning systems to maintain precise location during operations.
- **Unmanned Surface Vehicles (USVs) with LARS**: Increasingly, autonomous or remotely operated surface vehicles are being used as mobile platforms for underwater vehicle launch and recovery. USVs offer the advantage of reducing human risk and can operate in challenging conditions with advanced station-keeping capabilities (Figure 10.8) (Thales, 2020).
- **Hybrid Vessel Systems**: These systems combine the features of manned and unmanned vessels to offer a flexible and cost-effective platform for underwater operations (Figure 10.9).

FIGURE 10.7 Ramp-type LARS.

Source: **Okeanus.**
Notes: **https://okeanus.com/products/launch-and-recovery-systems/ramp-style-lars**

10.3.3 Autonomous and Remote LARS

Autonomous systems aim to minimise or eliminate human intervention during launch and recovery:

- **Automated LARS on USVs**: In these systems, an unmanned surface vessel is equipped with an automated launch and recovery mechanism. For example, a system may use acoustic positioning and remote-controlled winches to synchronise the retrieval of an AUV (Sarda & Dhanak, 2018).
- **Robotic Docking Stations**: Some systems use fixed or mobile docking stations that allow underwater vehicles to 'dock' autonomously for battery recharging, data transfer, or maintenance (Figure 10.10). These are particularly valuable in long-duration missions where frequent retrieval to the surface is not feasible (Figure 10.11).

The development of automated launch and recovery systems integrated into unmanned surface vehicles (USVs) involves the following aspects:

- **Autonomous Coordination**: The USV is equipped with a top-mounted winch and telescopic arm system that can lower an AUV into the water while the USV is in motion.

FIGURE 10.8 SAND (Surface Advanced Naval Drone) USV of Fincantieri operating as mother platform for AUVs.

Notes: **https://www.navalnews.com/naval-news/2023/07/fincantieri-goes-unmanned-with-sand-usv/**

- **Acoustic Positioning**: The recovery process utilises acoustic homing signals to guide the AUV towards a taut line suspended from the USV.
- **Reduced Human Risk**: By automating the most dangerous aspects of the launch and recovery process, the system reduces the need for personnel to be exposed on deck during adverse conditions.
- **Demonstrated Feasibility**: Modelling and simulation using tools like OrcaFlex have shown that the system can function effectively in low to moderate sea states, with ongoing research aimed at extending its operational envelope.

10.3.4 Custom and Modular LARS Solutions

Custom solutions are developed to meet specific mission requirements or to integrate with unusual platforms. On the other hand, modular systems are designed with interchangeable components that can be reconfigured to handle different vehicles or operational conditions. For example, a modular LARS might use interchangeable winch assemblies that can be swapped out based on the weight and size of the vehicle (Figure 10.12).

FIGURE 10.9 Henriksen LARS demonstrated for HUGIN AUV from an unmanned boat (RIB).

Source: **H.Henriksen AS.**
Notes: **Hydro International, 2020; https://sea-technology.com/henriksen-lars-demo-with-hugin**

FIGURE 10.10 Towed docking system for autonomously docking the Explorer AUV.

Source: **ISE Canada.**
Notes: **ISE Ltd, 2025; https://ise.bc.ca/product/dock/**

Some missions require customised solutions, such as deep-water recovery of heavy vehicles or recovery in highly dynamic environments. Systems like NOV-BLM's LARS for offshore energy applications exemplify the development of high-capacity, robust systems designed for extreme conditions (NOV, 2025; IHC, 2025).

FIGURE 10.11 ARV-I AUV with docking station for wireless battery charging and transferring collected data.

Source: **Boxfish Robotics.**
Notes: **https://www.boxfishrobotics.com/boxmedia/arv-i/**

FIGURE 10.12 AUV Amogh being deployed from the harbour for demonstration.

Source: **Larsen & Toubro.**

In military applications, LARS may be integrated with other mission systems—for example, using on-board sensors, acoustic homing, and guidance algorithms to enable rapid recovery during high-stress operations. These systems are often designed for stealth and rapid redeployment.

10.4 DESIGN APPROACHES FOR LARS

The design of an effective LARS is a multidisciplinary challenge that involves mechanical, hydrodynamic, control, and materials engineering. The following subsections describe common design approaches and key considerations.

10.4.1 Mechanical Design Considerations

10.4.1.1 Structural Components

The primary structural components of a LARS include cranes, winches, A-frames, and telescopic arms. These components must be designed to withstand both static loads (the weight of the underwater vehicle) and dynamic loads (forces generated by waves, vessel motion, and rapid acceleration/deceleration). Engineers often use finite element analysis (FEA) to simulate stress distribution, fatigue, and potential failure modes. Materials chosen for these components are typically high-strength, corrosion-resistant alloys or composites that can operate in saltwater environments. Manufacturing techniques such as precision machining, welding under controlled conditions, and advanced composite layup are employed to produce components that meet the stringent performance requirements of LARS.

10.4.1.2 Winch and Cable Systems

Winches are critical for controlling the descent and ascent of the vehicle. Modern winch systems incorporate variable-speed controls and are designed for rapid deployment and recovery while providing sufficient braking to manage dynamic loads. The cable used for ROV umbilicals, or for lowering AUVs, must balance flexibility with high tensile strength and resistance to corrosion and abrasion. Some systems also include tether management systems (TMS) to minimise drag and entanglement during operations.

10.4.1.3 Shock Absorption and Heave Compensation

Shock absorbers and heave compensation systems are designed to mitigate the effects of vessel motion on the vehicle during launch and recovery. Systems may use hydraulic dampers, spring-loaded mechanisms, or active control algorithms that adjust winch speeds in real time. In some cases, telescopic arms are integrated with shock-absorbing elements to smooth the transition through the turbulent splash zone.

10.4.1.4 Hydrodynamic Design

Hydrodynamic performance is critical in ensuring the smooth passage of the vehicle through the water. Designers must account for:

- **Drag Reduction**: Minimising hydrodynamic drag is essential for energy efficiency and for preventing undue strain on mechanical components. Streamlined shapes and the careful selection of materials can help reduce drag.
- **Flow-Induced Vibrations**: Vibration due to turbulent water flow can lead to fatigue and wear on both the LARS and the underwater vehicle. CFD simulations help engineers predict and mitigate these effects.
- **Alignment and Stability**: Ensuring that the vehicle remains aligned during deployment is crucial. This may involve designing guiding structures, such as rails or funnels, that direct the vehicle into the correct orientation before it is fully in the water.

10.4.2 Control and Automation

Automation in LARS design is increasingly vital to ensure safe and efficient operations. Key aspects include the following:

- **Sensor Integration**: Modern LARS incorporate an array of sensors, including accelerometers, gyroscopes, GPS, and acoustic position transducers. These sensors provide real-time feedback on vessel motion and vehicle position, allowing the control system to adjust operations dynamically.
- **Automated Control Algorithms**: Advanced control algorithms enable the system to compensate for external disturbances automatically. These algorithms are responsible for adjusting winch speeds, aligning the vehicle, and ensuring a smooth transition between air and water.
- **Human–Machine Interface (HMI)**: Even in automated systems, a robust HMI is necessary for monitoring and manual override in emergency situations. Clear display of sensor data, load status, and system health are essential components of modern LARS control panels.

10.4.3 Integration with Vessel Systems

The LARS must be integrated into the support vessel's structure and systems. This involves ensuring that the mounting points, power supplies, control interfaces, and communication networks are all compatible. In many cases, LARS are designed as modular systems that can be retrofitted to existing vessels, minimising the need for extensive vessel modifications. Dynamic positioning (DP) systems on the vessel also play a crucial role by maintaining the precise location required for safe launch and recovery operations.

10.5 EXAMPLES OF LARS SOLUTIONS

Several customised LARS solutions have been developed to meet the specialised requirements of different applications. Some examples illustrate how different approaches are used in practice.

The REMUS Launch and Recovery System developed by the Woods Hole Oceanographic Institution (WHOI) has been used for hundreds of launches and recoveries in various sea states (up to sea state 5). Key features include a tilting A-Frame with Integrated Heave Compensation to gently lower the vehicle into the water. Sensors integrated into the system help align the vehicle during launch and recovery.

NOV-BLM has developed a launch and recovery system specifically designed for offshore energy applications (IHC, 2025; NOV, 2025). Their system features a lifting capacity from 3 to 50 tons, making it suitable for large underwater vehicles, crew boats, and even small submarines. The operation is fully electric and has an adjustable and shock-absorbent cradle for recovery.

The HUGIN AUV was deployed and recovered using a purpose-designed launch and recovery ramp integrated with the AUV's storage container. This allows flexibility in storage, transportation, and deployment from the AUV's own container from any suitable ship with minimal modifications (Polar AUV Guide: Deployment and Recovery, n.d.).

Military operations require robust and often stealthy LARS solutions for the rapid deployment of unmanned underwater systems. An example is provided by L3Harris, which has demonstrated an integrated system for launching and recovering AUVs from an underway submarine (L3Harris, 2023). For Torpedo Tube Launch and Recovery (TTL&R), a compact recovery mechanism is integrated into a submarine's torpedo tube, allowing an AUV to be launched and later recovered without exposing the host vessel. The system was successfully demonstrated in tests where an AUV was deployed to perform reconnaissance missions while the submarine remained covert.

Although primarily designed for salvage operations, systems like the Flyaway Deep Ocean Salvage System (FADOSS) share many design principles with LARS for underwater vehicles.[1]

Table 10.1 provides a summary of several representative LARS solutions, their capacities, example applications, and key features.

10.6 MANIPULATOR SYSTEMS

Manipulator systems refer to robotic arms or tooling devices. These are critical components for underwater vehicles tasked with intervention, inspection, and sample collection beneath the water surface. UUVs with manipulators are often called Underwater Vehicle Manipulator Systems (UVMS). For the observation of their surroundings, they are usually accompanied by additional equipment comprising one or more cameras and spotlights mounted on the base underwater vehicle and/or on the manipulator itself (Sivcev et al., 2018).

ROVs, traditionally tethered and human-operated, depend extensively on manipulators for complex physical interactions in subsea environments. While AUVs focus principally on autonomous data gathering, limited specialised AUVs (Intervention-AUVs) incorporate manipulator arms to expand operational capability (Figure 10.13).

TABLE 10.1
Summary of Solutions of LARS

Name	Lifting/Deploying Capacity	Example Application	Key Features
WHOI REMUS LARS	Suitable for small AUVs (REMUS series)	Oceanographic research	A-frame system; integrated heave compensation; acoustic guidance; proven reliability up to sea state 5
NOV-BLM's Offshore LARS	3 to 50 tons	Offshore energy operations; unmanned crew boats, small submarines	Fully electric; adjustable, shock-absorbent cradle; modular design; can operate in high sea states; robust for heavy loads
USV-Based Automated LARS	Suitable for small AUVs (REMUS 100)	Autonomous deployment and recovery from USVs in moderate sea states	Automated winch and telescopic arm; acoustic positioning for precise recovery; minimises human exposure; integration with USVs
L3Harris Torpedo Tube LARS	Compact system for submarines	Covert AUV deployment from submarines	Integrated in torpedo tubes; rapid and stealthy launch/recovery
FADOSS Salvage System	Up to 20+ tons	Salvage operations; deep-water recovery of aircraft, submersibles	Modular, airlift-deployable; ship motion compensator; specialised winch and storage reel; rapid installation on vessels of opportunity

This section presents an overview of manipulator systems integrated with underwater vehicles, examining their design, functionality, control approaches, and applications. This material is drawn from industry reports and research literature, such as the comprehensive reviews by Sivčev et al (Sivcev et al., 2018). and Aldhaheri et al (Aldhaheri et al., 2022).

10.6.1 Role and Importance

Manipulator systems equip underwater vehicles with physical interaction capabilities, enabling inspection, intervention, and retrieval.

- **Inspection Tasks**: Close visual and sensor-based examination of seabed infrastructure, marine habitats, and shipwrecks.

FIGURE 10.13 REACH-X ROV demonstrating its high-dexterity 7 DOF arm with 3 kg lift capacity.

Source: **Reach Robotics.**
Notes: **https://reachrobotics.com/products/manipulators/**

- **Intervention Tasks**: Operation of valves, cutting cables, attaching lifting hooks, and assembling subsea equipment.
- **Sampling and Retrieval**: Collecting biological specimens, sediment cores, or artefact recovery.

In ROVs, manipulator arms have long been essential; their need emerged clearly in offshore oil and gas tasks, enabling maintenance and repair at depths beyond diver reach. Modern ROVs may carry multiple manipulators tailored for strength, dexterity, and tooling compatibility.

Advanced AUVs increasingly incorporate manipulators for autonomous intervention. Intervention autonomous underwater vehicles (I-AUVs) not only have to cope with the technical challenges associated with traditional manipulation tasks, but also do so while currents and waves perturb the stability of the vehicle, and low-light, turbid water conditions impede perceiving the surroundings. An example is shown in Figure 10.14.

FIGURE 10.14 TRIDENT UUV (2010–2012) demonstrated tele-operated or semi-autonomous manipulation tasks underwater.

Notes: **Aldhaheri et al., 2022; https://www.researchgate.net/figure/Examples-of-underwater-robot-manipulators_fig 1_357734227**

A 'fully teleoperated' I-AUV achieves a manipulation task by relying entirely on the teleoperators' skills (while being constrained by the need for low-latency communication) for navigating and acquiring the target object. On the other hand, 'fully autonomous' manipulation means that the I-AUV is able to detect an object, plan, grasp, and act on it without the intervention of a user. Some of the projects that have demonstrated such autonomous manipulation capabilities are Semi Autonomous Underwater Vehicle for Intervention Mission (SAUVIM) in 2005, TRIDENT, and Twin Robots for Cooperative Underwater Intervention Missions (TWINBOT) (Aldhaheri et al., 2022).

10.7 MANIPULATOR SYSTEMS DESIGN

10.7.1 Mechanical Design

Underwater manipulators are typically robotic arms designed to replicate human arm motions with multiple degrees of freedom (DOF). The design involves joints, links, specialised end-effector, appropriate materials, and housing.

Most of the underwater manipulation missions fall into the category of object retrieval. In some cases, the manipulation relies on the DOFs provided by the mobile base to achieve the mission. A series of rigid links are connected via revolute joints, allowing complex trajectories. Most manipulators have between 5 and 7 DOF for requisite dexterity.

The size of underwater manipulators is described by a parameter called 'Reach' which represents the length of the whole manipulator kinematic chain. Along with the range of motion of joints, it determines the size of the manipulator workspace, i.e., a set of points that can be reached by its end-effector. The reach of existing underwater manipulators ranges from 0.5 m for the grabber manipulators, up to 2.4 m for heavy-duty manipulators. Maximum wrist torque, which underwater manipulators are capable of producing, ranges from 8 Nm to 250 Nm, while the lifting (carrying) capacity ranges from 5 kg to 500 kg.

The end-effector refers to specialised tooling at the end of the manipulator arm, which may be grippers, cutters, or sensor mounts, as per mission needs. Underwater manipulators often come with interchangeable grippers whose purpose ranges from grabbing different objects to cutting an abandoned underwater net, for example. The most common gripper type is the two-parallel finger gripper. The most common actuators used to eject the gripping force are hydraulic, but electrical actuators have also been used for this purpose.

For underwater manipulation, replicating tactile sensing has proven to be a hard task. The most commonly used finger sensors are based on waterproof actuators that drive the fingers by means of a cable-based transmission. Force/torque sensors are typically integrated on each for grasp and force control (Aldhaheri et al., 2022).

Underwater manipulator weight (in air) is between 6 kg and 150 kg; however, their weight in water is more important, as it determines the buoyancy needed on the base vehicle in order to compensate for the manipulator. The weight and size are very important factors as they are directly responsible for the amount of dynamic coupling introduced between the manipulator and the underwater robot on which it is mounted, and can thus influence the performance of the whole system (Sivcev et al., 2018).

All components must withstand high pressures and corrosion. Lightweight alloys and corrosion-resistant materials are used.

10.7.2 Actuation and Power

Power for manipulators on ROVs is typically supplied via the tether; AUV manipulators require internally stored energy, with efficient power use critical to mission duration.

Actuation is achieved through electric motors, hydraulics or a combination of both. Brushless DC motors provide precision and ease of control for smaller manipulators, common in AUVs. Hydraulic actuators are used to power heavy-duty manipulators on work-class ROVs, and are vital for tasks requiring substantial lifting force. Combinations of electric and hydraulic actuators can optimise performance and efficiency.

Hydraulic actuators present do not require mechanical components (e.g., gears and levers) to exert force, thus offering a more efficient power-to-weight ratio. On the other hand, electric actuated manipulators, although heavier, provide more precise motion and force/torque control (Xu et al., 2025).

Leading commercial manufacturers of hydraulic underwater manipulators are Kraft Telerobotics, Schilling Robotics, Cybernetix, Hydro-lek, among others. Examples of models include the Schilling Titan-4 and Cybernetix Maestro. X Eca Robotics and Ansaldo are major commercial manufacturers of electric underwater manipulators. The MARIS is an electric 7-DOF manipulator by Ansaldo, rated for 6,000 m depth, used in many AUV and ROV systems (Sivcev et al., 2018).

There are also other modalities, such as pneumatic and tendon-driven actuators, that are popular for underwater soft-robotic applications, which will be introduced in the next section.

10.7.3 Manipulator Control Systems

Manipulator control for UUVs is complicated by underwater dynamics, communication delays (especially for AUVs), and uncertain environments. Hydrodynamic forces can be large and hence have a significant effect on the dynamic performance of underwater manipulation systems (McLain & Rock, 1998). The major challenges in manipulator control include the following aspects:

- **Kinematics and Inverse Kinematics**: Computing joint angles to achieve a target position and orientation (end-effector pose).
- **Force and Torque Control**: Managing interaction forces with objects or the environment to avoid damage.
- **Coordination with Vehicle Motion**: Mobile manipulators require coordinated control of the manipulator and the vehicle body to achieve precise operations, further complicated by environmental disturbances (Kumar, Rastogi, & Gupta, 2016).

10.7.4 Control Architectures

In contrast to conventional dry robotic manipulation on land, in underwater settings, sensory input is affected by limited accessibility and the inherent dynamics of the environment. These environmental changes result in flickering, insufficient illumination, refraction of the sensory input, and intermittent control.

Therefore, manipulator trajectory tracking control has to be backed up with the implementation of interaction force-torque control with the environment. Methodologies should consider that slight deviations of the end-effector from the planned trajectory could cause the manipulator to either lose contact with the surface (i.e., drop the object, presumably located with great difficulty) or generate large interaction forces that lead to damage (and mission failure).

Low-Level Control refers to direct motor or hydraulic valve commands with feedback for stability. Higher-Level Planning refers to path and trajectory planning, collision avoidance, and task sequencing (which we have seen in detail earlier). Advanced methods include weighted least-norm solutions, task-priority control, or learning-based control. Weighted least-norm approaches manage redundant DOF to optimise for secondary tasks or avoid joint limits. Learning-based adaptive control algorithms could progressively improve manipulation precision and robustness (Lane et al., 1997).

For ROVs, manipulators are tele-operated; joystick or haptic interfaces convey user commands. ROV operations, therefore, depend heavily on skilled personnel. Pilot proficiency affects operational efficiency, equipment safety, and mission success rate.

Autonomous control of manipulators is still an emerging area in advanced AUV systems. Autonomous operation would rely on sensor-driven decision-making and model-based control to execute complex tasks without direct human intervention. Challenges in underwater communication, sensing, data acquisition, and data processing, which also govern their mission efficiency, would continue to limit the efficacy of AUVs in autonomous manipulation as well.

10.8 TYPICAL MANIPULATOR SYSTEMS

10.8.1 ROV Manipulators

- **Work-Class Manipulators**: Robust hydraulic arms capable of heavy lifts (>50kg), multiple integrated tools, multi-DOF control, ergonomically designed for extended teleoperation (Robotics, 2022).
- **Light Class Manipulators**: Smaller electric arms common on observation and inspection ROVs, focusing on agility over brute strength.

10.8.2 AUV Manipulators

- **MARIS (Ansaldo) 7-DOF**: One of the early successful AUV manipulators, enclosed in pressure housing, integrated via the modular approach.
- **SAUVIM**: AUV mounted manipulator with coordinated vehicle-arm control demonstrating limited autonomous intervention.

- **Hybrid Systems**: Emerging architectures integrate morphing hulls or multifunctional tools shifting between survey mode (AUV) and manipulation mode (ROV), blurring domain boundaries.

10.9 ACTUATORS FOR SOFT ROBOTICS

In general, any actuator for underwater robotics enables movement and manipulation. In addition to the traditional connotation of actuators with control surfaces, or thrusters, they can also provide the force and control required for underwater movement (locomotion), object manipulation, and task performance.

Traditional UUVs that use thrusters or water jets for propulsion can be referred to as 'rigid body' robotics, where the form of the vehicle does not alter to provide motion or manipulation. A bio-inspired manipulator design is shown in Figure 10.15 (Sverdrup-Thygeson et al., 2016).

10.9.1 Soft Robotics

In contrast to 'rigid robotics', soft robotics uses fluidic actuators or 'smart' actuators to produce deformation of the robot's body (or an appendage), thus generating movement or manipulation. Soft robots are challenging to actuate due to their numerous degrees of freedom, resulting from the extensive deformation of the soft materials they are composed of. This leads to underactuated systems that are more difficult to control. Underwater soft robotic systems add extra DOF to the robotic manipulators.

As an example, a soft robot can have tendon wires, anchored at multiple points within its body. Electric motors, like servos, drive these cables by applying tension, causing the connected soft material to deform and produce various motions or shape

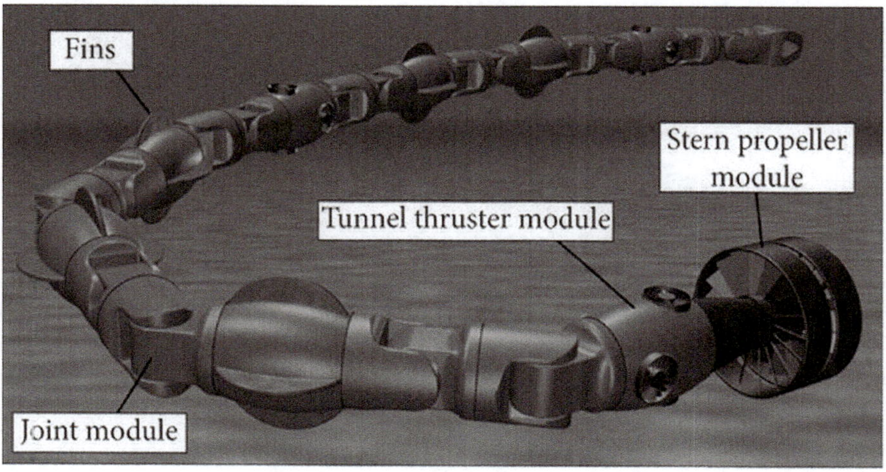

FIGURE 10.15 Design of a bio-inspired underwater manipulator.

Notes: **Neira et al., 2021; https://onlinelibrary.wiley.com/doi/10.1155/2021/5542920**

changes. Thus, the silicone arm of a bio-inspired octopus, powered by cables, can mimic the grasping and crawling abilities of an actual octopus's tentacles, (Hasib et al., 2025).

Another method of actuation involves various types of smart materials. Smart materials respond differently to external thermal or electric stimuli, changing their stiffness or causing deformation. Electroactive polymers (EAPs) deform in response to electric stimuli. One type of EAP, the dielectric elastomer actuator (DEA), consists of two compliant electrodes that compress when a high voltage is applied, producing motion. Compression force can generate motion, as demonstrated in biomimicry of jellyfish. Electrically driven actuators can thus be realised through mechanical/servo-motors, piezoelectric materials, or Dielectric Elastomers (Poka, Ludovico, & Manara, 2025).

Soft actuators are like artificial muscles to achieve motion and deformation. They can be powered electrically or driven by pressure, light, thermal changes, etc. Shape Memory Alloys (SMA) are thermal 'soft actuators', which can use shape memory polymers, neurostimulation, or liquid crystal elastomers. Fluidic elastomer actuators (FEAs) are another class of soft actuators that primarily rely on fluid pressure (Hasib et al., 2025).

An example of a sensor-less underwater snake robot is shown in Figure 10.16, which has the potential of performing tasks such as wrapping around a rod and then turning on/off a lever underwater (Kakogawa, Rajendrakumar, & Iwasaki, 2024). A survey of diverse types of bio-inspired soft underwater robots has been summarised and listed in Table 10.2.

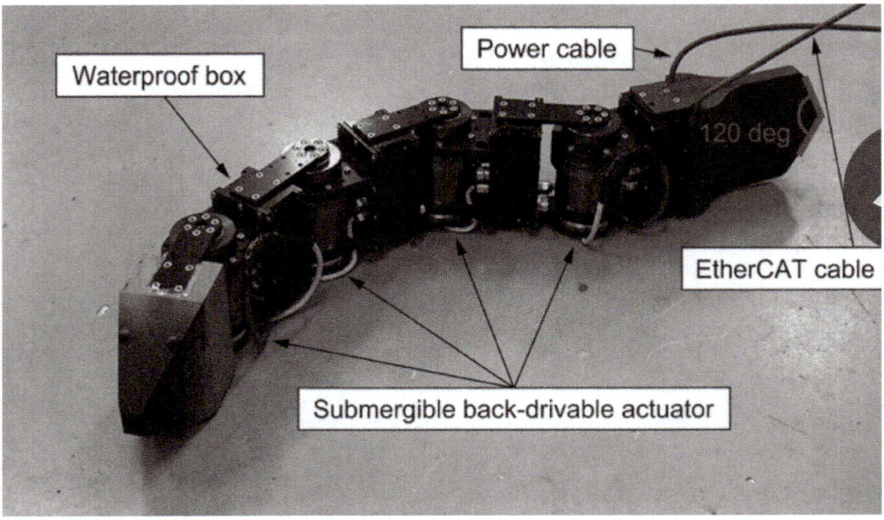

FIGURE 10.16 A bio-inspired underwater snake robot.

Notes: **Kakogawa, Rajendrakumar, & Iwasaki, 2024; https://www.fujipress.jp/jrm/rb/robot003600061458/**

TABLE 10.2
Soft Actuation for Bio-inspired Movements of Underwater Robots

Robot Type	Actuation/Materials Type	Movement Type	Characteristics	Year
Robo-jelly	Shape memory alloy (SMA) composite actuators	Through propulsion	Robo-jelly with a segmented bell and a passive flap structure	2011
Starfish, soft underwater robot	Servo-driven Tendon wires	Through propulsion	Made of silicone foam and actuated with four tendons	2021
Soft cuttlefish robot	Dielectric elastomer actuators (DEA)	Through jet propulsion	Soft artificial muscle-driven	2018
Brittle star-inspired 'Patrick'	Tendon wires that are Servo-driven	Through crawling	Five flexible legs actuated by 20 shape-memory alloy	2020
Octopus robot	Motor-driven cables or SMA	Through crawling	Eight radially extending arms with a central body housing processing units	2015
Underwater bionic crab soft robot	DEA	Controlled movements; crawls laterally	Multidirectional: Lateral crawling, forward underwater movement, ascent and descent	2023
Untethered soft robotic fish	Thermoelectric pneumatic actuators	Operates as a collective swarm	Mimic locomotion and untethered delivery of information using bidirectional comm.	2022
Underwater soft robot	Pneumatic-Hydraulic hybrid soft actuator—driving fin (DF)	Swimming and crawling	Two fibre-embedded soft actuators	2023
Jellyfish-like robot—SoJel	Flexible thermoplastic polyurethane material; Uses NiTi Flexinol	Swims with two different bell diameter skirt sizes	A bell-like system with eight segments arranged symmetrically, Flexinol	2023
FinRay robotic jellyfish	FinRay actuators driven with servo motors	Through propulsion	Central mantle with four tentacles, which are moved by a rod and a servo	2020
Octopus arm	Motor-driven cable or SMA springs	Elongation, shortening, and bending	SMA springs are used transversely, and cables are used longitudinally, to achieve actuator action	2012
Morphing limb amphibious turtle robo	Variable stiffness material pneumatic actuators	Drag-induced swimming or walking	Adapts to an unstructured environment	2019

(*Continued*)

TABLE 10.2 (CONTINUED)
Soft Actuation for Bio-inspired Movements of Underwater Robots

Robot Type	Actuation/Materials Type	Movement Type	Characteristics	Year
Starfish robot	SMA wires	Movement through propulsion	Equipped with five SMA tentacles	2016
Underwater soft octobot	SMA tentacles: simultaneous thrust and actuation	Swimming	Two SMA muscles in each tentacle to generate back thrust, increasing the frequency of activation	2022
Underwater snake robot	Hybrid actuation of stiff propulsion modules and soft joints	Multimodal swimming	Soft joints and stiff modules linked in series. Rectilinear motion in both directions; yaw, pitch, and spiral rising motion	2022
Miniature robot	Piezoelectric ceramics (PZT)	Swim, twist, straight drive; hover, sink, float	Actions quicker than current small underwater robots	2023
Mackerel	SMA	Swimming, undulatory motion	Swims well over most of the body's longitudinal orientations	2022
Octopus	Printed moulds, pour silica gel	Bipedal walking	Constructed from two trained soft arms	2022
Hydraulic soft robotic arm	Soft elastomer and single-chamber actuation	Manipulator	Uses nonlinear viscoelasticity of soft materials.	2022
Octopus	Soft silicon material	Grasping objects	Soft gripper to hold a moving turtle, irregular objects, scattered things, flat objects, etc.	2022
SoFi fish	Undulating hydraulic silicone elastomer tail	Swimming and turning motions	Two symmetric control surfaces for pitch control and diving	2022
Soft robotic fish	Antagonistic fast-PneuNet actuators	Swimming	Hyperelastic eutectic gallium–indium (eGaIn) embedded in silicone channels for strain sensing	2021
Jelly-Z	Twisted and coiled polymer (TCP)	Swimming	–	2023
Eel-like robot	Fluid Electrode DEA (FEDEA)	BCF undulation	Soft, actuated submersible device	2018
Octopus arm	Motor-driven cables actuation	Movement through crawling	Robot arm to grasp objects and move in a pushing-based manner	2011
RoboScallop	Fluidic elastomer actuators (FEAs)	Jet propulsion	Pulsating thrust via a single-DoF, reciprocating crank mechanism housed in a shell enclosure	2019

(Continued)

TABLE 10.2 (CONTINUED)
Soft Actuation for Bio-inspired Movements of Underwater Robots

Robot Type	Actuation/Materials Type	Movement Type	Characteristics	Year
Octobot	Chemical reaction-based FEAs	Swimming	Moulding and soft lithography used for the robot's body, microfluidic logic	2016
Starfish-like soft robot	Shape memory alloy	Movement through crawling	SMA actuators modelled on the deformation of two soft rays	2014
Jellyfish-inspired soft robot	DEA	Movement through propulsion	Waterproof untethered power source to facilitate propulsion	2019

Notes: Ibid.

10.10 CONCLUSION

We have seen that deployment and recovery of UUVs is a challenging part of operations, for which reliable LARS and their proper operation are essential. The types of LARS solutions and considerations in their selection and design have been discussed. Operator training and frequent operations are essential to gain proficiency in handling launch and recovery at sea under variable weather conditions. Along with the safety of handling personnel, the safety of an expensive UUV representing years of effort is also essential. Standardised solutions for launch and recovery are offered for established models of AUVs. In the case of a new development AUV, the design of the LARS should be given due consideration during the vehicle design process itself.

Manipulator systems extend the functional envelope of underwater vehicles beyond observation to active intervention and complex subsea work. Advances in mechanical design, actuation, and control facilitate increasingly capable, reliable manipulative functions on both AUVs and ROVs. Integration within modular, fault-tolerant architectures, supported by sophisticated control frameworks and emerging autonomy promises transformative capabilities in underwater exploration, industry, and defence.

Among underwater vehicles, manipulator control is relevant mainly for those ROVs that are intended for intervention or object recovery, which is primarily the case for the offshore oil and gas industry.

Soft actuators and soft robotics are propelling bio-inspired underwater robots, which are likely to attract greater interest in the years to come.

NOTE

1. https://www.navy.mil/Resources/Photo-Gallery/igphoto/2002536255/

REFERENCES

Aldhaheri, S., Masi, G. D., Pairet, E., & Ardon, P. (2022). Underwater Robot Manipulation: Advances, Challenges and Prospective Ventures. In *OCEANS-2022* (pp. 1–7). Chennai: IEEE. https://doi.org/10.48550/arXiv.2201.02954

Ballard, R. D., & Drew, C. (2021). *Into the Deep: A Memoir from the Man who Found Titanic.* National Geographic.

Christ, R. S. (2011). *The ROV Manual: A User Guide for Observation Class Remotely Operated Vehicles.* Netherlands: Elsevier Science.

Hasib, S. A., Gulzar, M. M., Oishy, S. R., Maaruf, M., Habib, S., & Shakoor, A. (2025). An Investigation of Innovative Strategies in Underwater Soft Robotics. *Engineering Science and Technology, 70.* https://doi.org/10.1016/j.jestch.2025.102123

Hydro International. (2020, January 16). This Is How an Unmanned Launch and Recovery System Operates. *Hydro International.* https://www.hydro-international.com/content/news/this-is-how-an-unmanned-launch-and-recovery-system-operates

IHC. (2025). *Launch and Recovery Systems (LARS).* IHC Offshore Energy. https://www.royalihc.com/offshore-energy/offshore-equipment/launch-and-recovery-systems

ISE Ltd. (2025). *Autonomous Docking: Discover Unlimited Range: Surfacing Optional.* International Submarine Engineering Ltd. https://ise.bc.ca/product/dock/

Kakogawa, A., Rajendrakumar, S., & Iwasaki, Y. (2024). Grasping of Cylindrical Structures Using an Underwater Snake Robot without Force/Torque Sensors and Actuator Waterproofing. *Journal of Robotics and Mechatronics, 36*(6), 1458–1467. https://doi.org/10.20965/jrm.2024.p1458

Kumar, S., Rastogi, V., & Gupta, P. (2016, December). Recent Developments in Modeling and Control of Underwater Robot Manipulator: A Review. *Indian Journal of Science and Technology, 9*(48). https://doi.org/10.17485/ijst/2016/v9i48/104371

L3Harris. (2023, July 03). Successful Launch and Recovery of Autonomous Underwater Vehicle from Underway Submarine. *L3 Harris.* https://www.l3harris.com/newsroom/editorial/2023/07/successful-launch-and-recovery-autonomous-underwater-vehicle-underway

Lane, D. M., Davies, J. B., Casalino, G., Bartolini, G., & Cannata, G. (1997, December). AMADEUS: Advanced Manipulation for Deep Underwater Sampling. *IEEE Robotics & Automation Magazine, 4*(4), 34–45. https://doi.org/10.1109/100.637804

McLain, T., & Rock, S. M. (1998, July). Development and Experimental Validation of an Underwater Manipulator Hydrodynamic Model. *The International Journal of Robotics Research, 17*(7). https://doi.org/10.1177/027836499801700705

Neira, J., Sequeiros, C., Huamani, R., Machaca, E., Fonseca, P., & Nina, W. (2021, January). Review on Unmanned Underwater Robotics, Structure Designs, Materials, Sensors, Actuators, and Navigation Control. *Journal of Robotics, 2021*(1). https://doi.org/10.1155/2021/5542920

NOV. (2025). Launch and Recovery System (LARS). *NOV.com.* Retrieved January 2025, from https://www.nov.com/products/launch-and-recovery-system-lars

Pantheeradiyil, A. (2024, October 7). How ROVs and Divers Are Transforming Underwater Inspections for Safety and Efficiency. *blueyerobotics.com.* https://www.blueyerobotics.com/blog/how-rovs-and-divers-are-transforming-underwater-inspections-for-safety-and

Poka, A., Ludovico, D., & Manara, F. (2025). Underwater Snake-Like Robots: A Review on Design, Actuation, and Modelling Methods. *International Journal of Advanced Manufacturing Technology, 139*(11–12), 5445–5460. https://doi.org/10.1007/s00170-025-16231-1

Polar AUV Guide: Deployment and Recovery. (n.d.). Society for Underwater Technology. Retrieved May 2025, from http://wsprdaemon.org/polar/deployment-and-recovery.html

Ray, A., Singh, S., & Seshadri, V. (2011, June 29–30). *Underwater Gliders - Force Multipliers for Naval Roles. Warship 2011: Naval Submarines and UUVs.* Bath, UK: Royal Institution of Naval Architects.

Robotics, R. (2022, August 3). Top Uses of Underwater ROV Manipulators. *Reach Robotics.* Retrieved January 26, 2025, from https://reachrobotics.com/blog/top-uses-for-underwater-manipulators/

Rumson, A. G. (2021). The Application of Fully Unmanned Robotic Systems for Inspection of Subsea Pipelines. *Ocean Engineering, 235.* https://doi.org/10.1016/j.oceaneng.2021.109214.

Sagar, M. M., Konara, M., Picard, N., & Park, K. (2025, February). State-of-the-Art Navigation Systems and Sensors for Unmanned Underwater Vehicles (UUVs). *Applied Mechanics, 6*(1). https://doi.org/10.3390/applmech6010010

Sarda, E. I., & Dhanak, M. R. (2018, September). Launch and Recovery of an AUV from a Station-Keeping USV. *IEEE Journal of Oceanic Engineering.* https://doi.org/https://doi.org/ 10.1109/JOE.2018.2867988

Sivcev, S., Coleman, J., Omerdic, E., Dooly, G., & Toal, D. (2018, September). Underwater Manipulators: A Review. *Ocean Engineering, 163*, 431–450. https://doi.org/10.1016/j.oceaneng.2018.06.018

Sverdrup-Thygeson, J., Kelasidi, E., Pettersen, K. Y., & T., G. J. (2016). The underwater swimming manipulator - A bio-inspired AUV. *IEEE/OES Autonomous Underwater Vehicles (AUV)* (pp. 387–395). Tokyo: IEEE.

Thales. (2020, November 26). *Thales to deliver the World's First Fully Integrated Unmanned Mine Countermeasures System for the Royal Navy and French "Marine Nationale.* Retrieved from thalesgroup.com: https://www.thalesgroup.com/en/group/journalist/press_release/thales-deliver-worlds-first-fully-integrated-unmanned-mine

WHOI. (2025). *Launch and Recovery Systems (LARS).* WHOI. Retrieved January 2025, from https://www.whoi.edu/what-we-do/explore/underwater-vehicles/auvs/remus/launch-and-recovery-systems-lars/

Wynn, R. B., Huvenne, V. A., Le Bas, T. P., Murton, B. J., Connelly, D. P., & Bett, B. J. (2014). Autonomous Underwater Vehicles (AUVs): Their Past, Present and Future Contributions to the Advancement of Marine Geoscience. *Marine Geology, 352*, 451–468. https://doi.org/10.1016/j.margeo.2014.03.012

Xu, Z., Zhang, Y., Chang, Z., Huang, B., & Bing, Y. (2025, December). Development of an Underwater Vehicle-Manipulator System Based on Delta Parallel Mechanism. *Journal of Marine Science and Engineering, 13*(12), 2361. https://doi.org/10.3390/jmse13122361

11 Testing and Trials

The transition of an Unmanned Underwater Vehicle (UUV) from a laboratory setting to its intended aquatic environment represents the definitive validation of its design and engineering integrity. The nature of subsurface operations is risky, where technical failures may result in the loss of high-value assets, damage to support infrastructure, or risks to personnel. A structured and phased evaluation strategy is therefore imperative (Griffiths, Millard, & Rogers, 2003). A thoughtfully planned sequence of tests and trials is essential for time-bound development. These need to be phased, and measurement parameters linked meaningfully to mission requirements.

Memoirs of UUV development and operations are filled with challenges faced and overcome during testing (Butler, 2018). This chapter outlines a comprehensive Experimental Test and Evaluation (ET&E) approach, designed to mitigate operational risks through incremental validation.

11.1 SIGNIFICANCE AND PHASES

Traditional Testing and Evaluation Master Plans (TEMP) often require modification to accommodate the probabilistic nature of autonomous underwater vehicles (AUVs). Thorough testing minimises these risks by ensuring that every subsystem, from propulsion and navigation to communication and power management, functions correctly under a range of conditions. The incremental approach to testing allows designers and engineers to validate assumptions, refine control algorithms, and optimise mechanical designs before exposing the vehicle to the full rigours of open water (Moore, Bohm, & Jensen, 2010). AUVs are complex systems and not necessarily suited to the mould of a traditional T&E Master Plan (TEMP). T&E must be flexible and opportunistic, while comprehensive enough to extract quantifiable measures of capability from a probabilistic system (Macias, 2008).

A robust testing regime serves several critical functions (Keane & Joiner, 2020):

- **Performance Verification**: Systematic trials generate quantitative data regarding velocity, endurance, manoeuvrability, and depth ratings, which are vital for calibrating control systems.
- **System Integration**: These vehicles require the seamless convergence of hardware, software, and communication protocols. Methods such as Hardware-in-Loop (HIL) testing are employed to verify these interfaces before sea deployment.
- **Regulatory Compliance**: Rigorous testing provides the empirical evidence necessary to satisfy commercial and military certification standards.

DOI: 10.1201/9781003737476-11

The evaluation of UUVs typically progresses through three distinct environments, from Hardware-in-Loop testing to harbour trials to sea trials. Each successive phase of trials increases in complexity and environmental realism.

11.2 HARDWARE-IN-LOOP (HIL) TESTING

HIL testing is conducted within a laboratory or simulated environment. It involves interfacing physical components such as controllers and actuators with high-fidelity simulations of marine dynamics. The simulation could mimic the dynamic behaviour of the ocean, including hydrodynamic forces, vessel motions, and sensor feedback.

Simulation environments allow for rapid testing of various scenarios, enabling engineers to adjust control parameters, refine algorithms, and evaluate system responses under a wide range of conditions (Dynautics, 2021). However, technical challenges often arise from simulation inaccuracies, such as the failure to model transient turbulence, which can lead to discrepancies in steering responses compared to real-world conditions.

HIL testing is crucial for verifying that the real hardware components interact correctly with the vehicle's software. This is particularly important for ensuring that sensor data is accurately interpreted and that actuator commands produce the expected results (Christ, 2011). In a typical HIL setup for underwater vehicles, the following components are integrated:

- **Control Hardware**: This includes the vehicle's central processing unit, motor controllers, and sensor interfaces.
- **Simulated Environment**: Using software such as MATLAB/Simulink, OrcaFlex, or specialised marine simulation tools, the dynamic behaviour of the underwater environment is modelled.
- **Real-Time Data Exchange**: Communication between the physical hardware and the simulation occurs in real time. The hardware receives simulated sensor inputs (e.g., simulated depth, speed, and orientation) and sends actuator commands back to the simulation.
- **Feedback and Monitoring**: Data acquisition systems and real-time monitoring dashboards are used to observe system behaviour, analyse performance, and identify any discrepancies between expected and actual responses.

11.3 HARBOUR TRIALS

Harbour trials provide a transition phase in sheltered waters to evaluate system integration with launch and recovery systems (LARS). These trials focus on:

- **Static and Dynamic Testing**: Initial deployment verifies power systems and sensor readings, followed by low-speed manoeuvring to assess navigation accuracy.

- **Safety Protocols**: Operators execute emergency drills, including manual overrides and abort sequences, to ensure fail-safe mechanisms are operational.
- **Environmental Sensitivity**: Real-water conditions may reveal unforeseen issues, such as buoyancy systems reacting erratically to minor temperature fluctuations.

While harbour trials are conducted in relatively calm conditions, they still present a range of challenges. During harbour trials for a new AUV design, the vehicle's buoyancy control system may turn out to be sensitive to slight temperature changes in the harbour water. This could result in erratic depth control during low-speed manoeuvres and may need to be addressed by recalibrating the buoyancy system and improving the insulation of critical components.

11.4 SEA TRIALS

This final phase subjects the vehicle to the rigours of the open ocean, including variable currents, waves, and complex underwater terrain. The primary objective is to validate environmental resilience and long-term system reliability. These trials progress from initial 'shakedown runs' (Figure 11.1) to full-mission simulations that replicate the vehicle's intended operational profile. A notable challenge during sea

FIGURE 11.1 Indian high endurance AUV undergoing lake trials.

Source: DRDO.
Notes: Menon, 2025; https://www.navalnews.com/naval-news/2025/04/indias-high-endurance-auv-program-undergoes-developmental-trials/

trials is acoustic communication disruption, where multipath interference or ambient noise can delay command execution (Xiang Liu, 2025; Zhang, 2020).

11.4.1 Objectives of Sea Trials

Key Objectives of Sea Trials are listed below.

- **Operational Performance Evaluation**: Sea trials assess the vehicle's overall performance in terms of speed, endurance, manoeuvrability, depth capability, and sensor functionality under real-world conditions.
- **Environmental Resilience**: These trials test the vehicle's ability to withstand and operate in adverse weather conditions, high waves, strong currents, and other unpredictable ocean dynamics.
- **System Reliability and Robustness**: Continuous operation in the open sea helps identify potential reliability issues—such as component wear, unexpected interactions between subsystems, or degradation of sensor performance.
- **Safety and Emergency Procedures**: Sea trials validate the full range of safety protocols, including abort sequences, emergency recovery, and fail-safe mechanisms. Real-world scenarios such as loss of power or communication are simulated to assess the robustness of these procedures.
- **Data Collection and Mission Simulation**: For research and commercial applications, sea trials are critical for collecting operational data. This data is used to refine operational parameters, improve mission planning, and optimise sensor performance.

11.4.2 Stages of Sea Trials

Sea trials are usually conducted in a series of stages:

- **Shakedown Runs**: Initial short-duration deployments are conducted to verify that the vehicle can leave the support vessel, execute basic manoeuvres, and return safely. These runs often focus on verifying the functionality of critical systems such as propulsion, control, and communications.
- **Incremental Mission Profiles**: Once the shakedown phase is complete, the vehicle is tested under increasingly challenging conditions. This may involve deeper dives, extended missions, and more complex manoeuvres such as obstacle avoidance and automated path following.
- **Full-Mission Simulations**: The final phase of sea trials replicates the full operational mission, including long-duration deployments, data collection, and interaction with other systems (such as docking stations or unmanned surface vehicles).
- **Post-trial Analysis**: Data collected during sea trials is analysed to assess performance, identify failures, and recommend modifications. This

analysis is crucial for certification, regulatory approval, and iterative design improvements.

11.4.3 Sea Trial Challenges

Sea trials present the most rigorous test conditions and thus reveal many challenges that may not be apparent during HIL or harbour trials:

- **Environmental Uncertainty**: Open-sea conditions are unpredictable. Sudden weather changes, unexpected currents, and large waves can cause the vehicle to behave erratically or even damage sensitive components.
- **Integration of Systems**: The complexity of integrating mechanical, electronic, and software systems becomes more apparent in a real-world setting. For example, the interaction between dynamic positioning systems on the support vessel and the vehicle's control system can be challenging to synchronise.
- **Communication Disruptions**: Underwater communication remains one of the most challenging aspects of sea trials. Acoustic modems may suffer from multipath interference, low bandwidth, or dropouts due to environmental noise. This can lead to delays in command execution and reduced situational awareness.
- **Operational Fatigue**: Extended missions can reveal issues such as battery degradation, thermal management challenges, or gradual drift in sensor calibration. These factors require robust design and redundancy.
- **Safety Incidents**: Despite extensive pre-trial testing, unexpected incidents can occur during sea trials. For instance, a vehicle might encounter an unanticipated obstacle or experience a temporary loss of control. Such incidents, while undesirable, provide invaluable data for improving system resilience and emergency procedures.

11.5 TYPICAL CHALLENGES

Testing underwater vehicles is a complex, iterative process. Typical challenges specific to each stage of testing have been mentioned. Some common issues likely to be faced across various stages of the testing process of UUVs are listed below for ready reference.

11.5.1 Simulation vs. Reality Discrepancies

One of the most common challenges in HIL testing is the gap between simulated conditions and real-world dynamics. For example, a simulation may not accurately capture the chaotic behaviour of ocean currents or the rapid fluctuations in sea state. This can lead to control algorithms that perform well in the lab but fail under actual conditions. Continuous refinement of simulation models using data from harbour and sea trials is necessary to minimise these discrepancies.

11.5.2 Sensor and Communication Issues

Underwater vehicles rely heavily on sensors for navigation, control, and data collection. During testing, issues such as sensor drift, calibration errors, and data latency can significantly impact performance. Acoustic communication, which is often used in underwater trials, is particularly prone to interference from ambient noise and multipath effects. Such issues can lead to delays in command execution or even temporary loss of communication, complicating the testing process.

11.5.3 Mechanical and Structural Failures

The marine environment is inherently harsh, with saltwater, pressure, and dynamic loads all contributing to wear and tear on mechanical components. During sea trials, even minor design flaws in the vehicle's structure or in the launch and recovery system can lead to failures. Problems such as cable fatigue, hydraulic system leaks, and joint wear are not uncommon, and they often require on-the-fly modifications and post-trial repairs.

11.5.4 Integration and Interface Mismatches

Underwater vehicles are composed of many interdependent subsystems. Ensuring that these subsystems interact seamlessly is a major challenge. Integration issues may arise from incompatible communication protocols, mismatched electrical interfaces, or differing response times between subsystems. Such problems often become apparent only during integrated testing in harbour or sea trials.

11.5.5 Environmental Uncertainties

The variability of ocean conditions is one of the most unpredictable factors affecting testing. Sudden changes in weather, unexpected tidal currents, or the presence of debris can all impact the performance of an underwater vehicle during sea trials. Operators must be prepared to handle these uncertainties, often requiring backup procedures and contingency plans.

11.5.6 Human Factors and Training

Despite the increasing automation of underwater vehicles, human operators still play a critical role in managing trials. Operator errors, whether due to insufficient training or misinterpretation of sensor data, can lead to incidents during testing. Continuous training, simulation exercises, and the development of robust human–machine interfaces are essential to mitigate these risks.

11.6 PLANNING SEQUENCE OF TRIALS

Each AUV mission profile has unique requirements with distinct operational priorities; however, a number of underlying capability factors or measures of effectiveness have been identified (e.g., navigational accuracy, sensor effectiveness, and clearance rates). These universal factors form the basis of most AUV capabilities, such that Experimental Test and Evaluation (ET&E) can be standardised to an extent, thereby enabling system comparison to determine the suitability of different AUVs for various operating profiles (Kass, 2015).

The main goals of T&E are to reduce risk and provide information for acquisition decisions or answer capability questions posed by the end-user. The two key phases in constructing the framework are the Design of Experiments (DoE) approach to establishing key factors and outputs, and the development of a test plan (Antony, 2014). Once combined, such a framework can test for the technical and operational measurements required to determine whether the AUV capability achieves its intended purpose, and to answer typical questions posed by decision-makers (potential customers or funding agencies) (Gooding, 2001).

Deriving a Process Flow (PF) diagram outlines the typical operational flow for AUV deployments, which is critical in understanding where technical and operational measurements are appropriate. This can form the backbone of the ET&E framework (Figure 11.2).

11.6.1 Example of MCM AUV

Consider an AUV developed for the Mine Countermeasures (MCM) role involving search. The AUV can be considered as part of systems-of-systems for the MCM

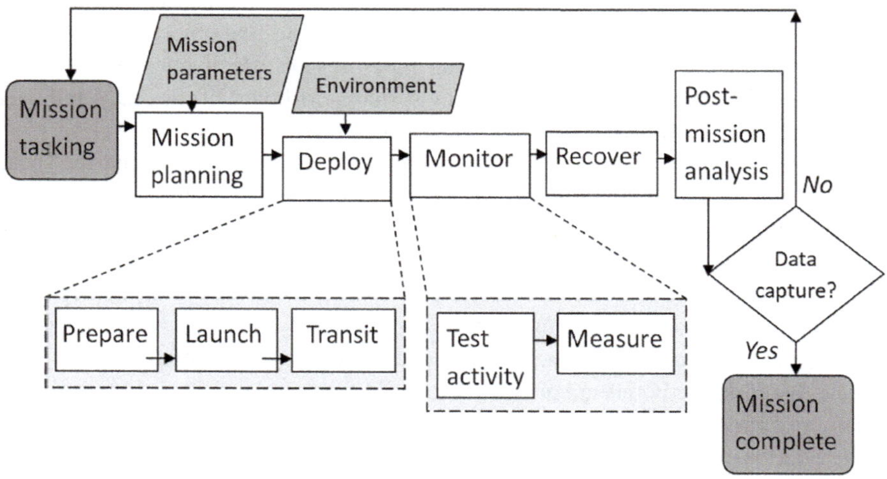

FIGURE 11.2 Process flow diagram for AUV operational test and evaluation.

Notes: **Based on Keane and Joiner (2020).**

TABLE 11.1
Overview of Test Activities for an MCM AUV

Activity	Purpose of Test	Overview
Payload	Effectiveness of sonar and camera payloads	Payload effectiveness is tested over a dedicated range of targets. These targets could be mine-like objects or unclassified targets such as crayfish pots. Effectiveness is measured as a probability of detection over numerous runs.
Communications	Range and reliability of different communication methods	Building a database of communication performance during ET&E is useful as it contributes to further development of collaborative autonomy developments
Vehicle dynamics	Characterise AUV dynamics by different manoeuvres	Vehicle dynamics during AUV missions, when subject to environmental factors, are critical to the quality of payload data.
MCM search	Characterise AUV ability to detect mine like objects	The primary mission capability of the AUV is evaluated once sufficient results from other activities have been derived. This activity aims to measure tactical outputs such as clearance rate, probability of detection, coverage achieved, and mission completion rates. These are tested in a range of operating environments.

Notes: Keane and Joiner (2020)

effect. For example, in the context of a most basic MCM Critical Operating Issue (COI), ‘Can the AUV capability search and detect a mine-like object?’, the requirements development includes decomposing measures of effectiveness (MOE) and Measures of Performance (MOP) for many systems off-board of the AUV and over longer-term MCM mission timeframes (Gooding, 2001).

Four core activities were derived for this mine countermeasures scenario. These are detailed in Table 11.1, and all necessary test results can be derived from variants of these activities.

The user effectiveness (COI) might be evaluated as the answer to a question, such as, ‘Can the AUV detect a Mine Like Object’? To answer this, further questions are required to properly understand the capability, such as sensor accuracy and confidence of classification. The payload effectiveness, communications, and vehicle dynamics activities are necessary to build the foundational understanding of AUV capability before determining the effectiveness of the COI. This order is due to the relationship between navigational accuracy, sensor effectiveness, and vehicle dynamics.

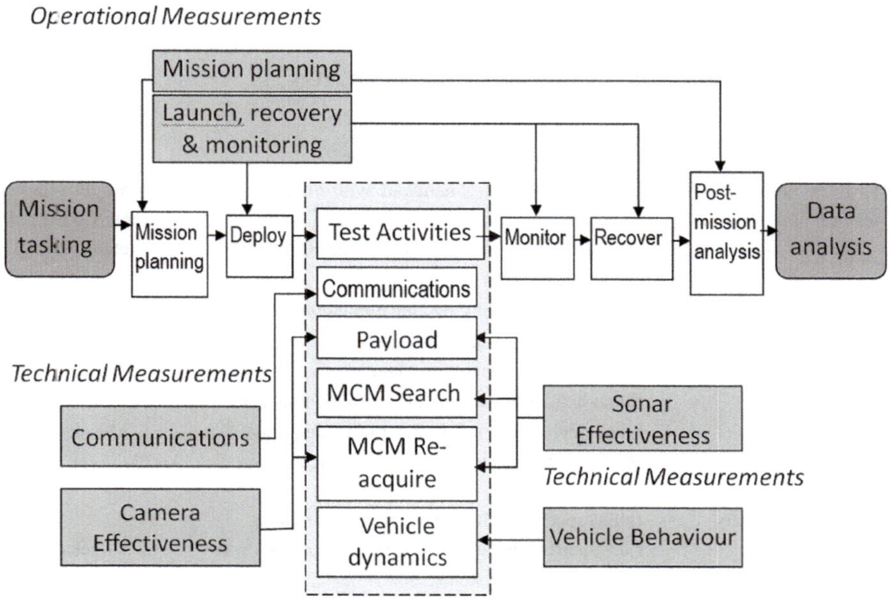

FIGURE 11.3 Experimental test and evaluation framework for an AUV.

Notes: **Based on Keane and Joiner (2020).**

11.6.2 Test and Evaluation Plan

The ‘Design of Experiments’ methodology and first-principles approach discussed above can be used to draw up an Experimental Test and Evaluation (ET&E) framework that combines the Process Flow (PF) and operational requirements; an AUV ET&E framework is shown in Figure 11.3.

Table 11.2 summarises an idealised sequence with approximate durations for each phase. Note that actual durations may vary based on the vehicle’s complexity, environmental conditions, and specific mission requirements. The durations indicated below (total about 20 to 30 weeks) are idealised estimates for a medium-complexity underwater vehicle.

11.7 CONCLUSION

Underwater vehicle testing and trials are an indispensable part of the development process, ensuring that vehicles can operate reliably in the harsh and unpredictable marine environment. Each phase serves a specific purpose, from early risk mitigation and subsystem integration to comprehensive operational performance evaluation. This progression reflects the strategy of gradually increasing complexity and exposure to risk. Each phase builds upon the successes and lessons of the previous stage; hence, any shortcuts may be disastrous. Future advancements in big data analytics and remote testing platforms are expected to further enhance the efficiency and safety of these evaluation protocols.

TABLE 11.2
Idealised Sequence for Testing and Trials of an AUV

Phase	Activity Description	Approximate Duration
Hardware in-loop (hil) testing	- Integration of control hardware with simulation environment. - Testing of sensor inputs, control algorithms, and actuator responses. - Iterative refinement based on simulation data.	4–6 weeks (as per complexity and iteration cycles)
Laboratory component testing	- Bench testing of individual subsystems (e.g., propulsion, buoyancy, communication). - Stress and fatigue testing of critical components.	2–4 weeks
Controlled environment (harbour) trials—preliminary	- Static deployment in a harbour or testing basin. - Verification of basic deployment and recovery procedures. - Initial operator training and system calibration.	1–2 weeks
Controlled environment (harbour) trials—dynamic	- Low-speed manoeuvring and integration tests with launch/recovery systems. - Simulation of emergency procedures and manual override tests.	2–3 weeks
Incremental sea trials—shakedown	- Short-duration sea trials near the support vessel. - Verification of system performance under real-water conditions. - Collection of baseline performance data.	1–2 weeks
Incremental sea trials—extended missions	- Progressive increase in mission complexity (deeper dives, longer durations, complex manoeuvres). - Testing of full-mission profiles and emergency procedures under variable conditions.	3–4 weeks
Full-scale sea trials and final certification	- Comprehensive deployment in operational sea conditions. - Performance evaluation over extended missions (full-endurance, high-speed manoeuvres, deep dives). - Final data collection for certification and regulatory compliance.	4–6 weeks
Post-trial analysis and feedback	- Detailed review and analysis of all trial data. - Identification of design improvements and operational adjustments. - Documentation and training updates.	2–3 weeks (overlapping with trial phases)

REFERENCES

Antony, J. (2014). *Design of Experiments for Engineers and Scientists*. London: Elsevier.

Butler, B. (2018). *Into the Labyrinth: The Making of a Modern-Day Theseus*. Canada: Bigfoot Press.

Christ, R. S. (2011). *The ROV Manual: A User Guide for Observation Class Remotely Operated Vehicles*. Netherlands: Elsevier Science.

Dynautics. (2021, June 15). Modelling, Testing & Sea Trials – All without Leaving The Lab. *Dynautics*. https://www.dynautics.com/modelling-testing-and-sea-trials-all-without-leaving-the-lab/

Gooding, T. R. (2001). *A Framework for Evaluating Advanced Search Concepts for Multiple Autonomous Underwater Vehicle (AUV) Mine Countermeasures (MCM)*. Masters Thesis, Massachusetts Institute of Technology. Retrieved December 2023, from https://www.semanticscholar.org/paper/A-framework-for-evaluating-advanced-search-conceptsGooding/7be3b3f3b5c3a6c7bac73fa5b981a33de8096174

Griffiths, G., Millard, N., & Rogers, R. (2003). Logistics, Risks and Procedures Concerning Autonomous Underwater Vehicles. In *Technology and Applications of Autonomous Underwater Vehicles* (pp. 279–293). CRC Press. https://doi.org/10.1201/9780203522301.ch16

Kass, R. (2015). Twenty-one Parameters for Rigorous, Robust, and Realistic Operational Testing. *ITEA Journal, 36*, 121–138.

Keane, J., & Joiner, K. F. (2020). Experimental Test and Evaluation of Autonomous Underwater Vehicles. *Australian Journal of Multi-Disciplinary Engineering, 16*(6), 1–13. https://doi.org/10.1080/14488388.2020.1788228

Macias, F. (2008). The Test and Evaluation of Unmanned and Autonomous Systems. *ITEA Journal*, *29*, 388–395.

Menon, A. K. (2025, April 20). India's High Endurance AUV Program Undergoes Developmental Trials. *Naval News*. https://www.navalnews.com/naval-news/2025/04/indias-high-endurance-auv-program-undergoes-developmental-trials/

Moore, S., Bohm, H., & Jensen, V. (2010). *Underwater Robotics: Science, Design and Fabrication*. Monterey, CA: Marine Advanced Technology Education Center.

Xiang Liu, S. X. (2025). Innovative Strategy and Practice of Using Underwater Robot for Marine Cable Inspection and Operation and Maintenance. *Cognitive Robotics*. https://doi.org/10.1016/j.rineng.2021.100201

Zhang, H. et al. (2020). Subsea Pipeline Leak Inspection by Autonomous Underwater Vehicle. *Applied Ocean Research*. https://doi.org/10.1016/j.apor.2020.102321

12 Development Case Studies

Design and development of an engineering product transcends the efforts of a single individual or a single firm. Durable technological solutions require an entire ecosystem that connects needs with technological means, engineering ability, trial facilities, infrastructure, and support for through-life sustenance.

This chapter draws upon the experiences of UUVs in terms of product development to highlight the sequence and journey of their development, trials and induction. This would assist all stakeholders in the UUV ecosystem to better fathom the challenges involved.

We start by examining the chronological development of UUVs worldwide, followed by UUV developments in India. This is followed by case studies of the most successful UUVs in the world: the REMUS, the HUGIN, and the Bluefin. An overview is provided of US and Chinese developments in this field, and development timelines of typical UUVs are compared.

12.1 DEVELOPMENT OF UUVS

12.1.1 1950s–1970s: Pioneering Concepts

The post-World War II surge in oceanography, defence research, and Arctic exploration set the stage for the development of robotic undersea platforms.

The United States saw its first operational AUV, the Special Purpose Underwater Research Vehicle (SPURV), emerge from the University of Washington's Applied Physics Laboratory in 1957. Tasked with Arctic research, SPURV's untethered design made it a precursor to today's autonomous vehicles—capable of collecting vital data on ocean currents, temperature, diffusion, and submarine wakes, and reaching impressive depths of over 3,000 m. In the Soviet Union, parallel research was underway by the 1970s, producing AUV designs like 'SCAT', 'L1', and 'L2' for mapping and sensor development (Exploring NUSTEM, n.d.).

- **1957**: The United States launches the Special Purpose Underwater Research Vehicle (SPURV), one of the first AUVs, for oceanographic studies.
- **1960s**: The concept of underwater gliders emerges, with patents and experimental designs such as Concept Whisper (1960).
- **1966**: The US Navy debuts the Cable-Controlled Underwater Recovery Vehicle (CURV), enabling tethered deep-ocean salvage operations and setting early ROV industry standards.

 DOI: 10.1201/9781003737476-12

- **1970s**: Soviet programmes introduce early AUVs like SCAT, L1, and L2 for hydrographic data and sensor testing.

12.1.2 1980s–1990s: Breakthrough Prototypes

The 1970s and 1980s witnessed the refinement of ROVs (Remotely Operated Vehicles), notably characterised by human pilots controlling the machines via tethers from surface vessels. These platforms revolutionised deep-sea recovery, inspection, and salvage. They became instrumental in offshore oil/gas operations, underwater infrastructure inspections, and even scientific sample collection far beyond the reach of human divers (Blidberg, 2001).

Technological advances in microprocessors, batteries, and control systems enabled newer, more robust ROVs and ushered in the first wave of 'hybrid' platforms capable of limited autonomy. In 1983, the Autonomous and Remote Controlled Submarine (ARCS), equipped with cutting-edge processors from International Submarine Engineering, marked a critical testbed in vehicle control and endurance.

In the early 1990s, the MIT AUV Lab developed a vehicle series known as the Odyssey class. These were roughly 21 inches in diameter and about 2 m long. They were designed to dive as deep as 6,000 m but be relatively affordable and easy to deploy. These vehicles supported many science missions, including under-ice work in the Arctic and oceanography in the Antarctic. A key programme supported by these vehicles, sponsored by the Office of Naval Research (ONR), was the Autonomous Ocean Sampling Network (AOSN). Experiments in seafloor mapping and mine hunting were also conducted (Manley, 2019).

Underwater Gliders are a class of AUVs (autonomous underwater vehicles) that use buoyancy-driven propulsion and were developed for oceanographic studies in the mid-1990s.

By the 1990s, UUVs shifted from experimental test-beds to mission-oriented assets. Some of the milestones are summarised here:

- Work-class ROVs entered wide usage for undertaking inspection, construction, and maintenance of subsea oil and gas infrastructure, as well as salvage and science. Industry leaders included Perry Slingsby and Oceaneering.
- **1983**: The ARCS (Autonomous and Remote Controlled Submarine) demonstrated the fusion of remote control and emerging autonomy.
- **Late 1980s**: The UK develops Autosub, with the first model operating from 1996 for seabed mapping and under-ice science; subsequent models enable 6,000 m depth and multi-month endurance.
- **1990s**: The REMUS series (Woods Hole Oceanographic Institution) and Canada's Explorer (International Submarine Engineering) were introduced. REMUS was for naval mine countermeasures and hydrography, while Explorer was for under-ice navigation and modular commercial/scientific work.
- **1992**: The University of Tokyo tested ALBAC, an experimental underwater glider. Subsequent developments (late 1990s) of buoyancy-driven gliders

took place in the US at Webb Research, University of Washington and Scripps Institute of Oceanography.

- **1997**: Bluefin Robotics was founded to transition the developments in underwater robotics at MIT's AUV lab to industry, the first of many commercial AUV manufacturers to follow.

12.1.3 2000s: Production and Demonstration

The 21st century has seen UUVs move from specialised experiments to widely deployed operational systems. With a surge in funding, all stakeholders (the defence, industrial, and scientific communities) pursued innovations in autonomy, payload integration, and network communication.

The software 'Mission Oriented Operating Suite' (MOOS) was launched at the MIT AUV Lab in the early 2000s. Other technical developments undertaken were experiments with early acoustic modems and Doppler velocity logs (DVLs). Payload systems were also integrated, such as digital side-scan sonars, sub-bottom profilers, and docking.

The first practical underwater gliders were introduced in the early 2000s, with three pioneering models: the Slocum Glider (Teledyne Webb Research), the Spray Glider (University of Washington), and the Seaglider (Scripps Institution of Oceanography) (Ray, Singh, & Seshadri, 2011).

In parallel, the US Navy and partners like the Woods Hole Oceanographic Institution (WHOI) developed and fielded the REMUS family of AUVs. Small, modular, and robust, these vehicles were soon adopted by navies worldwide for surveillance and mine countermeasure (MCM) operations. By the end of the decade, vehicles could be launched from submarines, patrol vessels, or the shore, expanding their operational scope and flexibility. The REMUS 600 and 6000 series enhanced performance for deep-sea MCM and intelligence, surveillance, and reconnaissance (ISR) roles in challenging environments, such as the Middle East and the Western Pacific.

Some of the milestones in this period included:

- **2003**: The first underwater gliders (Teledyne Webb Slocum, Spray, and Seaglider) saw widespread deployment, offering months-long endurance by using buoyancy-based propulsion as a low-power alternative to propellers or thrusters.
- **2009**: The Scarlet Knight (Rutgers) becomes the first underwater glider to cross the Atlantic in a single mission, covering over 7,300 miles in 221 days.
- **2012**: Bluefin-21 AUV from Bluefin Robotics debuted, supporting offshore survey, mine countermeasures.

12.1.4 Since 2010: Expanded Roles and Integration

The 2010s have brought in a new level of software-driven autonomy and cooperative mission execution. AUVs began to feature adaptive algorithms, advanced path planning, and real-time decision-making—enabling them to cope with dynamic and complex undersea environments. Artificial intelligence and machine learning now support autonomous mission re-tasking, robust sensor fusion, and predictive health management on board.

Autosub Long Range and Explorer set new benchmarks for Arctic and Antarctic science, supporting kilometre-scale under-ice missions. Japanese innovations like the Urashima AUV achieved impressive endurance milestones. Bluefin-21 and variants performed multi-hundred-hour, hundreds-of-kilometres-long endurance missions in defence and commercial roles.

Advanced work-class ROVs (Oceaneering, Triton, PSS) now support global energy, subsea construction, and environmental monitoring. Their capabilities include deep-water cable laying, pipeline inspection, and nuclear power plant maintenance. Work-class ROVs have seen further automation with improved AI, automation, sensors, and integration with AUV fleets.

Underwater gliders (Slocum, Spray, Seaglider, SeaExplorer) routinely perform months-long, transoceanic research and surveillance, enabling real-time environmental monitoring, hurricane research, and climate science (Stommel, 1989).

It is envisaged that fleets of UUVs linked by secure underwater communications and digital command chains would enable collaborative search, distributed mapping, and rapid data relay to command centres. International exercises and real-world demonstrations have tested the potential of such networks for MCM and anti-submarine warfare (ASW) missions. Large unmanned platforms (XLUUVs) like the Boeing Echo Voyager/Orca, UK's Cetus and Kongsberg's HUGIN Endurance exemplify the strategic push for larger, multi-mission, long-endurance uncrewed vehicles.

Countries including the US, Russia, China, Iran, North Korea, and Ukraine have fielded or tested armed UUVs, from explosive-laden kamikaze drones to nuclear-powered oceanic strike vehicles like Russia's Poseidon/Status-6, blurring lines between ISR and offensive weapons. The ongoing Russia-Ukraine war since 2022 has brought into focus the use of drones in warfare on land and at sea (Dickinson, 2025).

12.2 INDIAN DEVELOPMENTS

The National Institute of Oceanography (of CSIR) at Goa was the first in India to develop an Autonomous Underwater Vehicle (AUV). This was *Maya*, developed in 2006, which was used for oceanographic data collection (Elgar Desa, 2007). The *Maya* project was funded by CSIR-NIO and the Ministry of Communications & IT of the Government of India (Maurya, Morishita, & Pascoal, 2022) (Figure 12.1).

In 2009–2010, another CSIR lab, the Central Mechanical Engineering Research Institute (CMERI) at Durgapur, West Bengal, developed the 'AUV-150' along with IIT Kharagpur. This project was funded by the Department of Ocean Development of the Government, which later became the Ministry of Earth Sciences (MoES, 2024). The AUV-150 was demonstrated at Idukki lake (Kerala) and subsequently at sea in 2011 (Shome, 2010). A deeper-diving version, AUV-500, was also developed but not tested. The AUV-150 is shown in Figures 12.2 and 12.3.

FIGURE 12.1 NIO Goa team with India's first AUV Maya.

Source: **CSIR-NIO.**
Notes: **Maurya, Morishita, & Pascoal, 2022; https://www.nio.res.in/research/technologies/autonomous-underwater-vehicles-auv**

FIGURE 12.2 CMERIAUV-150 lake trials in 2010.

Source: **CSIR-CMERI.**

FIGURE 12.3 Sea trials of AUV-150 by CSIR-CMERI off Chennai in 2011.

Source: **CSIR-CMERI.**

The Naval Science and Technological Laboratory (NSTL) of DRDO located at Visakhapatnam developed a 'flatfish'-form AUV around 2011 as a technology demonstrator (Korulla, 2016). This was also successfully tested at sea (Figure 12.4) but was not inducted.

The Government of India's Department of Ocean Development (which became the Ministry of Earth Sciences (MoES) in 2006) had set up the National Institute of Ocean Technology (NIOT) at Chennai in 1993. NIOT has developed ROVs (ROSUB 6000 and PROVe) (Sathianarayanan, 2013), which have been used for various expeditions.

The technology of AUV 'Maya' was transferred by NIO Goa to M/s Larsen and Toubro (CSIR-NIO, 2023), who subsequently demonstrated (from 2014) a series of AUV designs, developed with foreign collaboration (Kullashri, 2022).

12.2.1 Current Maturity

As a result of the impetus through various ongoing national missions, underwater robotics has attracted start-up firms as well as established technology giants and shipyards (iDEX, 2025). Several UUV prototypes have successfully transitioned to production orders for industrial and defence applications (NMICPS, 2023; Ray, 2025).

Some of the AUVs, ASVs and ROVs developed in the country by various institutions/firms since 2015 are shown in Figure 12.5.

FIGURE 12.4 Sea trials of flatfish AUV by NSTL off Visakhapatnam in 2011–2012.

Source: **DRDO-NSTL.**

FIGURE 12.5 Examples of AUVs and ROVs developed in India.

Notes: **Ray, Underwater Robotics – Engineering and Applications, 2025.**

12.3 CASE STUDY: *REMUS* AUV

The REMUS series of autonomous underwater vehicles (AUVs), developed by Woods Hole Oceanographic Institution (WHOI) and Hydroid (now part of Huntington Ingalls Industries), has become a cornerstone technology in underwater robotics. REMUS AUVs are renowned for their modular, flexible design and have achieved exceptional market penetration, with over 700 units sold to 30 countries as of 2025 (EMR Claight, 2025).

12.3.1 Origins and Development

The REMUS (**R**emote **E**nvironmental **M**onitoring **U**nit**S**) AUV concept originated in 1993 at WHOI, with the first vehicle completed in 1995 for oceanographic data collection. Early models were pioneering in their autonomous capabilities compared to then-common tethered systems, using primitive algorithms and rudimentary recovery systems.

Christopher von Alt was the leader of the team that originally developed the REMUS AUV as part of the WHOI. In 2001, von Alt co-founded Hydroid Inc., an independent company to commercially manufacture, support, and further develop the REMUS systems. At peak strength, Hydroid had 87 employees (full- and part-time), with representatives for local sales and support in nearly 30 countries. In 2008, Hydroid was acquired by Kongsberg Gruppen, and in 2020 by Huntington Ingalls Industries (HII) (Hydroid Inc., 2012; Kongsberg, 2009).

In 2002, the United Kingdom purchased REMUS vehicles for mine countermeasure operations. In 2003, REMUS became the first AUV used by the US Navy in combat during Operation Iraqi Freedom (Figure 12.6), surveying the Strait of Hormuz and Port of Umm Qasr for naval mines (Fotheringham, 2021).

12.3.2 Design Features

REMUS vehicles feature reconfigurable sensor payloads and modular batteries, allowing deployment across various missions, depths, and durations. Smaller REMUS models use robust designs suited for coastal and littoral operations, while deep-water versions like REMUS 3000 and 6000 use titanium construction for resilience at great depths (WHOI, 2025).

Standard sensors include conductivity, temperature, side-scan sonar, multibeam sonar, 3D bathymetric imagers, and customised customer payloads for research, surveillance, or industrial applications (HII, 2025). Special adaptations include SharkCam and TurtleCam for animal tracking via acoustic tagging and video.

The various models of REMUS AUVs are summarised in Table 12.1. The REMUS 620, launched in 2022, offers battery endurance of up to 110 hours and can carry synthetic-aperture sonar, smaller UUVs or UAVs, and allows for submarine launch/recovery. The REMUS 300 complies with Modular Open Systems Architecture (MOSA) and shares its core electronics and software with the REMUS family, facilitating updates and rapid payload swaps (Naval Technology, 2023).

FIGURE 12.6 REMUS600 AUV launched for mine search and identification operations in 2018.

Source: **MCC America Henry/U.S. Navy.**
Notes: **Eckstein, 2023; https://www.defensenews.com/naval/2023/11/08/us-navy-tests-sub-launched-drones-while-industry-continues-designing/**

TABLE 12.1
Types of REMUS AUV

Model	Max Depth	Endurance	Use Case	Notable Features
REMUS 100	100 m	10 hrs	Coastal surveys, mine countermeasures	Compact, lightweight
REMUS 300	305 m	8–12 hrs	Mine countermeasures, ISR	Modular, advanced autonomy
REMUS 600	600 m	Up to 70 hrs	Versatile, deeper ocean missions	Easily reconfigurable
REMUS 3000	3,000 m	~22 hrs	Deep-water research	Titanium hull, advanced sensors
REMUS 6000	6,000 m	Up to 22 hrs	Ultra-deep exploration	Modular, long range

12.3.3 Market Performance

The REMUS series is most widely adopted among small AUVs, with the REMUS 100 alone accounting for 156 production units as early as 2007, making it the most successful production model in its class at the time (Newman, Westwood, & Westwood, 2007).

The wider unmanned underwater vehicle market is projected to reach $19.8 billion by 2034, with REMUS being a major commercial and military product (GMI, 2024). North America, spearheaded by US defence and commercial interests, leads in deployments, while the Asia-Pacific market is rapidly growing due to naval modernisation and marine mining needs.

The REMUS series is a standard in the US Navy's mine countermeasure fleet and features in the next-generation SUUV programme, with REMUS 300 chosen as the programme of record. These autonomous systems can operate independently or teamed with crewed platforms such as Virginia-class nuclear submarines, expanding operational reach (Naval Technology, 2023).

Nations such as Japan, New Zealand, and the UK have adopted REMUS for seafloor mapping, mine countermeasures, environmental monitoring, and hydrographic surveys.

12.4 CASE STUDY: *HUGIN* AUV

The Norwegian firm Kongsberg Maritime manufactures the HUGIN AUVs. These AUVs are renowned for their precision, reliability, adaptability, and global impact in both commercial and defence spheres.

12.4.1 Development

HUGIN development began in 1991, aimed at mapping Norway's challenging offshore environments with high accuracy. In 1995, the HUGIN project was created in collaboration between Kongsberg, the Norwegian Defence Research Establishment (FFI), Statoil, Norsk Undervannsintervensjon (NUI), and the Royal Norwegian Navy (Hofoss, 2021). HUGIN's design was intended from the start for dual application in military mine countermeasures and oil and gas pipeline inspection (Wennberg, 2023). Over the past three decades, continuous innovation has allowed the HUGIN series to meet increasingly demanding applications in hydrographic surveying, oil and gas exploration, scientific research, and defence operations (Kongsberg, 2009).

The HUGIN AUV family has evolved into one of the most successful commercial AUV platforms (Figure 12.7).

12.4.2 Design Features

HUGIN is recognised as a reliable and stable platform, supporting high-accuracy sensor payloads, fair endurance, and minimal noise interference. Current HUGIN models operate at depths of 3,000, 4,500, or 6,000 m, with HUGIN Endurance extending this to long-range, shore-to-shore missions.

The series features swappable batteries (Lithium-Polymer or custom cells), flexible payload bays, and containerised designs for rapid mobilisation. Over time,

FIGURE 12.7 The HUGIN range of AUVs (top to bottom: HUGIN Edge, HUGIN, HUGIN Superior, and HUGIN Endurance).

Source: **Kongsberg.**
Notes: **Kongsberg, 2025; https://www.kongsberg.com/discovery/autonomous-and-uncrewed-solutions/auv/**

innovations such as plug-and-play payload systems and advanced battery technology were introduced to enhance versatility and endurance (Oceaneering, 2025).

Aided inertial navigation systems and high-precision acoustic links (like Kongsberg's HiPAP) support precise navigation both autonomously and under supervisory control with real-time feedback. The design evolved to incorporate a robust Aided Inertial Navigation System (INS) fused with Doppler Velocity Log (DVL) measurements, GPS surface fixes, and terrain-aided navigation updates. This combination of techniques enables high positioning accuracy in extended underwater operations (Jalving et al., 2004).

Typical payloads include synthetic aperture sonar, multibeam echo sounders, sub-bottom profilers, cameras, laser profilers, and environmental sensors.

Data handling features include removable storage for rapid data offload and cloud-based integration for remote monitoring and mission planning.

Features of various models of HUGIN AUV are summarised in Table 12.2.

Current generation AUV models feature advances in shore-to-shore operation, cloud data offload, and secure, real-time mission updates. Containerised solutions for storage, handling, and launch and recovery also enable ease of integration on a mother ship.

12.4.3 Trials and Demonstrations

The design cycle for HUGIN AUVs is characterised by a continuous loop of concept validation, prototype development, sea trial testing, and software/hardware updates.

In trial scenarios, aiding sensors (such as the DVL and terrain navigation modules) were selectively disabled to demonstrate the system's resilience (Jalving, Vestgård, & Storkersen, 2002). In another trial, a HUGIN 1000 vehicle maintained

TABLE 12.2
Types of HUGIN AUV

Model	Max Depth	Endurance	Key Usage
HUGIN I	600 m	~6 hrs initially	Pipeline survey, R&D
HUGIN 1000/1500	1,000–1,500 m	Up to 24 hrs	Survey, mine countermeasures, oil
HUGIN Edge	1,000 m	24 hrs	Deployment from USVs
HUGIN 3000	3,000 m	>60 hrs	Deep-water commercial
HUGIN 4500/6000	4,500–6,000 m	26+ hrs (4 knots)	Deep ocean mapping, search & rescue
HUGIN Endurance	6,000 m	15 days, 2,200 km	Long-range, shore-to-shore ops
HUGIN Superior	6,000 m	Enhanced, multi-mission	Highest-end scientific, military

navigation accuracy within 5 m over a 7-hour transit in challenging environments (Hagen, Anonsen, & Mandt, 2010).

A record-setting mission of the 'HUGIN Endurance' was carried out between September 2023 and March 2024 in a fully autonomous sortie (Marine Technology News, 2024). The 8-ton AUV operated without any human intervention or external navigation support for the 'multi-week' mission (Kongsberg, 2024). The vehicle operated at depths ranging from 50 m to 3,400 m and covered 1,200 nautical miles. Its mission included imaging and bathymetry over 36 square nautical miles using synthetic aperture sonar (SAS) and laser profiling. Operating without external fixes, the vehicle returned with a position error of approximately 0.02% of the total distance travelled.

12.4.4 Market Performance

Over 100 HUGIN units have been delivered worldwide, serving governments, navies, oil and gas companies, and scientific organisations. HUGIN is used for pipeline surveying, underwater infrastructure inspection, resource exploration, environmental monitoring, and salvage operations. HUGIN technology is exported globally, with rapid expansion in Asia-Pacific and Europe, addressing both commercial (oil and gas) and military needs(Skopljak, 2022).

Multiple NATO and allied navies employ HUGIN AUVs for mine countermeasures, ISR, and anti-submarine warfare, with demonstrated success during numerous operational deployments. At least 12 navies currently operate HUGIN AUVs. Recent contracts include a Defence Innovation Unit (DIU) award for the US Navy's large displacement UUV programme (Marine Technology News, 2024).

12.5 CASE STUDY: *BLUEFIN* AUV

The Bluefin series of AUVs, now part of General Dynamics Mission Systems, are modular, deep-rated platform that is extensively used for naval mine countermeasures, deep-sea search, offshore energy, and environmental monitoring, enjoying robust global sales and market reputation (Figure 12.8).

12.5.1 Origins and Development

Bluefin Robotics was founded in 1997 by engineers from MIT's AUV Laboratory, notably Dr James Bellingham and Frank van Mierlo, to commercialise the advanced Odyssey-class AUV technology developed at MIT. Early market analysis revealed strong demand in offshore energy services and defence. The company quickly transitioned from academia to industry, opening a dedicated production facility in 1999. Bluefin established itself through key contracts with the US Navy and federal agencies, contributing novel battery technology and modular designs to address deployments at extreme ocean depths and long durations (Sea Grant, n.d.).

The design was driven by the aim to provide a versatile survey and reconnaissance tool that could operate independently for extended periods (approximately 25 hours at a cruising speed of 3 knots) and be reconfigured rapidly to meet different mission requirements (Lundquist, 2023).

FIGURE 12.8 Bluefin-21 AUV being lowered for operation.

Source: **US Navy.**
Notes: **By Mass Communication Specialist 1st Class Peter D. Blair/U.S. Navy, Public Domain, https://commons.wikimedia.org/w/index.php?curid=32341642**

12.5.2 Design Features

Bluefin AUVs are constructed around interchangeable payload and battery modules. This architecture enables the rapid adaptation of vehicles for marine survey, archaeology, salvage, mine detection, oceanography, and other applications.

The pressure-tolerant lithium-polymer batteries, developed with SBIR and Navy contracts, offer both portability and remarkable endurance, powering long-duration missions down to 6,000 m and supporting an expanded suite of geophysical, chemical, magnetic, and imaging sensors. Bluefin's UN-certified pressure-tolerant battery design allows global transport and rapid emergency deployment, supporting multiple payload types for flexible mission response (SBIR-STTR, 2020).

Bluefin AUVs are generally torpedo-shaped for hydrodynamic efficiency and stability, with robust materials (carbon fibre, stainless steel) to withstand pressures at full ocean depth (General Dynamics, 2024).

All models employ a combination of inertial navigation systems (INS), Doppler Velocity Logs (DVL), and USBL (ultra-short baseline) acoustic positioning for high-accuracy georeferenced mapping. The Bluefin-9 features redundant communications and emergency navigation systems for safe recovery and reliability in critical missions.

Advanced on-board processing delivers highly accurate bathymetric, magnetometric, and imaging data, with automatic post-mission data readout. Data Management features include mission-specific flash and SSD storage of up to several terabytes and on-board real-time data processing. These enable Bluefin vehicles to deliver immediate post-mission results for sonar, camera, and sensor outputs (General Dynamics, 2015).

Various models of Bluefin AUV are summarised in Table 12.3.

TABLE 12.3
Variants of Bluefin AUV

Model	Max Depth	Endurance	Key Use Cases	Notable Features
Bluefin-9	200 m	8 hrs	Rapid deployment, port security	Two-man portable, quick swap battery, side-scan sonar
Bluefin-12	200 m	12 hrs	Survey, inspection	Lightweight, modular, 32 cm diameter
Bluefin-21	4,500–5,000 m	25 hrs @ 3 knots	Deep-sea search, military ops	13.5 kWh battery, high modularity
Knifefish	4,500 m	16+ hrs	Mine countermeasures (US Navy)	Advanced sonar, AI classification. Derived from Bluefin-21

12.5.3 Applications

- **Naval Mine Countermeasures**: Bluefin-21 and its derivative Knifefish are standard platforms in US and allied navies for detection, classification, and neutralisation of marine mines, offering deep and shallow water flexibility. One notable field trial was during the US Navy's biennial Ice Exercise (ICEX), where a Bluefin-21 (named 'Macrura') was tested under Arctic ice conditions (Lundquist, 2021). These trials verified capabilities such as autonomous navigation, operation of side-scan sonar during survey, and communication protocols (acoustic and satellite) during the mission.
- **Search and Recovery**: Bluefin-21 came to global prominence in the search for Malaysia Airlines Flight 370 in 2014, demonstrating deep-water search capabilities and reliability (Taylor, 2014).
- **Commercial Survey**: The Bluefin-21 entered service in the early 2010s, with commercial operators (such as Phoenix International) and defence agencies (including the US Navy and other research organisations) adopting the system for routine missions. It is used for offshore pipeline inspection, cable route surveys, mineral prospecting, infrastructure inspection, and environmental monitoring by leading oil and gas and marine engineering firms.
- **Environmental Research**: Equipped for seafloor mapping, water column sampling, biodiversity studies, and archaeological surveys in partnership with research institutes and governments.

As a General Dynamics product line, Bluefin is among the market's top revenue generators for untethered AUVs, competing directly with Kongsberg, SAAB, and Teledyne. Bluefin AUVs are found in navies, commercial fleets, and research labs worldwide, with hundreds of vehicles delivered to institutions in North America, Europe, Asia, and Oceania.

12.6 US NAVY PROGRAMMES

Several policy documents have been issued over the past two decades by various agencies of the US government and military, outlining visions for induction and utiliation of unmanned systems in warfighting (US Department of Defense, 2018).

Key US Navy programmes for UUVs include a spectrum from small vehicles (man-portable) and low-endurance (of a few hours), to large and extra-large vehicles (about 50 tons) with high-endurance (missions of 3 to 6 weeks). A snapshot of the UUV systems roadmap of the US Navy is shown in Figure 12.9.

12.6.1 Programme Progress

Several of these have been developed and demonstrated as prototypes, while others have been discontinued (GAO, 2022).

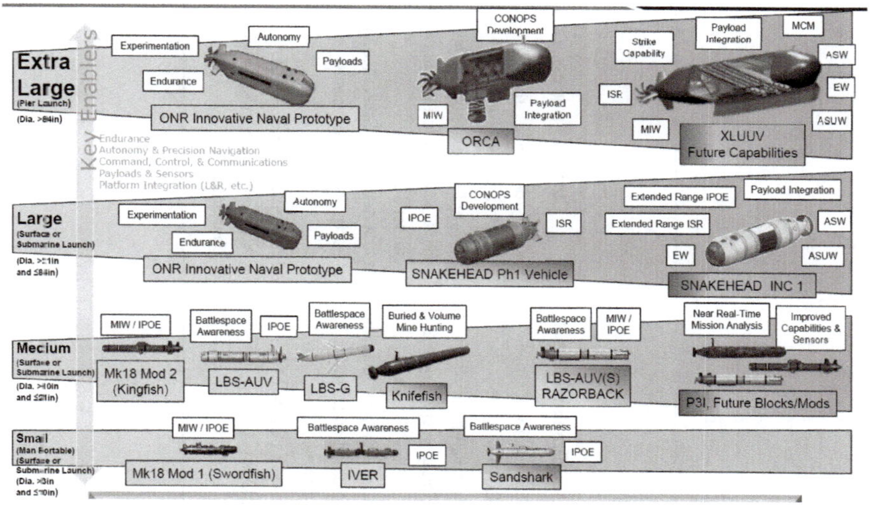

FIGURE 12.9 US Navy vision for unmanned systems.

Source: **US Navy, 2022.**
Notes: **https://www.navsea.navy.mil/Portals/103/Documents/Exhibits/SNA2019/UnmannedMaritimeSys-Small.pdf?ver=2019-01-15-165105-297 ; O'Rourke, 2025**

Table 12.4 summarises UUV categories as per the US Navy Roadmap and their roles, planned timelines, and progress.

The US DoD and US Navy roadmaps indicate the following core strategic goals:

- **Force Multiplication**: Enhance combat effectiveness without risking personnel.
- **Rapid Deployment**: Shorten acquisition cycles to rapidly field new capabilities.
- **Interoperability**: Ensure seamless integration between unmanned systems and existing manned platforms.
- **Cost Efficiency**: Emphasise modularity and mass production to reduce life-cycle costs.
- **Operational Flexibility**: Develop adaptable systems that perform multiple roles in contested environments.

12.7 CHINESE UNDERWATER SYSTEMS

Chinese AUV research and development began in the 1980s, with the Shenyang Institute of Automation (SIA) creating the country's first underwater vehicle, the HR-01, in 1981. Significant state investments, especially after the launch of the '863 Plan' in 1986 and its marine technology extensions following 1996, spurred rapid advances. Today, three principal research institutions anchor this field (Fedasiuk, 2021):

TABLE 12.4
US Navy UUV Roadmap Targets and Progress

Size Category	Example Program(s)	Primary Roles/Missions	Planned Timeline/Target	Current Progress
Small	REMUS 100, M3V	ISR, mine countermeasures, rapid environmental assessment	Fielded; operational since 2010–2015	Widely deployed; iterative upgrades ongoing
Medium	REMUS 300, Kingfish (Mk 18 Mod 2)	ISR, mine detection, moderate-range missions	Prototypes validated; transition expected by 2023	Selected as next-generation SUUV; operational trials underway
Large	Orca (XLUUV), Snakehead	Extended ASW, covert payload delivery, deep-sea operations	Orca: First delivery December 2023; Snakehead: Under development	Orca delivered; Snakehead in early prototype phases
Extra large	LDUUV, XLUUV	Strategic mine deployment, long-endurance, covert missions	Procurement targeted for mid-2020s	Prototypes tested; XLUUV experiencing cost/schedule challenges

- **Shenyang Institute of Automation (SIA)**: Developed the Explorer series (deployed in 1994) and the advanced Sea-Whale 2000 and Qianlong deep-sea-capable models.
- **Harbin Engineering University (HEU)**: Responsible for the Smart Water series, Weilong (Microdragon), and other survey platforms tested extensively in the South China Sea.
- **CSIC (China Shipbuilding Industry Corporation) 701 and 710 Research Institutes**: Pioneered large AUV systems such as Haishen 6000 ('Poseidon'), which can dive to 6,000 m and serve multi-sensor missions.

12.7.1 Application Spectrum

Scientific Exploration: Deep-sea AUVs like Sea-Whale 2000 and Haishen 6000 contribute to China's oceanographic, geological, and climate research, mapping previously inaccessible regions of the deep ocean.

Defence: Large AUVs such as HSU-001 are deployed for seabed warfare, surveillance, and intelligence gathering; smaller models are used in anti-submarine, mine countermeasure, and security roles (Strachan, 2019). Two extra-large UUVs (XLUUV), HSU100 and AJX002, both about 15 m long, were displayed in the annual Chinese military parade in September 2025 (Chuanren, 2025).

Resource Survey: These vehicles are indispensable for oil and gas prospecting, subsea mining, infrastructure inspection, and cable/pipeline route optimisation, particularly in the South China Sea.

Commercial and Civil Uses: The growing private sector supplies autonomous vehicles for marine engineering, environmental monitoring, and search-and-rescue operations.

12.7.2 Technical Capabilities

From torpedo-shaped deep-water models like Qianlong to modular gliders such as Haiyi, Chinese AUVs employ pump-jet propulsion and high-performance electric/hydraulic thrusters, often engineered for low acoustic signatures and high efficiency in extreme conditions.

Pressure-tolerant Li-ion and hybrid battery systems extend operational times to days for deep surveys or persistent military missions. Standard and advanced payloads now include multibeam side-scan sonar, sub-bottom profilers, magnetometers, real-time imaging, black box search arrays, and environmental sensors for broad mission profiles. Navigation is through integrated inertial systems, Doppler velocity logs, GPS, and acoustic positioning. These enable underwater route planning and mapping, often enhanced by AI and machine learning for autonomous and cooperative missions.

Chinese AUVs have been developed under various national R&D programmes (such as the 863 and 912 Programmes), and examples are shown in Figure 12.10.

FIGURE 12.10 Snapshots of Chinese AUVs and ROVs

Source: **Jianguo Wu.**
Notes: **Jianguo, 2018.**

These include several notable classes, with salient characteristics listed in Tables 12.5.

12.7.3 Prospects

China accounts for the largest share of the regional AUV market, driven by state investment, aggressive offshore energy initiatives, and naval modernisation.

Sales of autonomous underwater vehicles in China are forecast at $744.9 million in 2024, with continued growth seen from new applications and export potential (TechSci Research, 2025). Defence spending, offshore energy, deep-sea mining, and environmental monitoring are core sectors that are driving this growth. Rapid advances in AI, materials, batteries, and modularity continue to fuel domestic innovation and competitiveness (Goldstein, 2021).

Emerging private firms are penetrating regional and global markets. Chinese AUVs are poised for further sophistication, with technological priorities in perception, navigation, cooperative multi-vehicle operation, and endurance.

China is also investing in underwater sensor networks to support both civilian and military applications. These networks aim to provide continuous, high-bandwidth communications between underwater vehicles and command centres. Development of AUVs, ROVs and manned submersibles are linked with the development of these underwater sensor networks (Na, 2025).

12.8 DEVELOPMENT TIMELINES

The approximate development timelines (from concept to operational induction) for a range of contemporary AUV/UUV systems are summarised in Table 12.6. These

TABLE 12.5
Key Features of Various Chinese AUVs

AUV Model	Length (m)	Weight (Tons)	Key Features/Payload
Explorer Series	4.4	2.2	Modular sensor suite including side-scan sonar, multibeam, sub-bottom profiler, cameras, acoustic communication
Sea-Whale 2000	~3	~0.2	AI-based power management, long endurance, multi-sensor oceanographic and environmental monitoring
Qianlong Series	3.5–4.6	~1.5	Advanced navigation, seabed mapping, multi-sensor scientific and resource prospecting payloads
HSU-001	5	3.0	Large displacement UUV, seabed warfare, military reconnaissance
Haishen 6000	7.62	3.0	Deep-sea search and rescue, high-resolution sonar, lights, acoustic beacon, hydroacoustic sensors
Haiyi Glider	2	0.07	Wave/buoyancy-powered glider, extremely low acoustic signature, environmental sensing
Smart Water	Varies	Varies	Multi-variant survey platforms, flexible payloads for education, ocean monitoring
CR-01	4.374	1.3	GPS-assisted nav, reprogrammable microprocessors, black box for recovery
CR-02	4.5	1.5	Multi-sonar obstacle avoidance, ocean mineral survey, mud penetration ability
Micro Dragon	~1.5–2.5	Small scale	Miniaturised AUV family. Confined space operations, rapid response
Wukong	2	1.3	Full ocean depth operations, water sampling missions, Ultra-deep exploration, strategic ISR

durations vary depending on project scope, funding, technological challenges, and operational requirements.

While development timelines can vary widely depending on the system's size, intended mission, and technological complexity, Table 12.6 shows that modern AUVs typically require between three and nine years from concept to induction. Trends such as rapid prototype testing (as seen with Ghost Shark) and hybrid operational modes (as with Qilin ARV) are helping to shorten these cycles and meet the growing demand for autonomous undersea capabilities in both defence and commercial sectors.

- **Bluefin-21**: Its timeline reflects an evolution from concept through field trials (including high-profile missions) to induction in commercial and defence operations.

TABLE 12.6
Development Timelines for Various AUVs

UUV Name	Developer (Country)	Start of Development	Prototype Trials	Operational Induction	Total Duration
Bluefin-21	Bluefin Robotics (USA)	~2008–2009	2012–2014	Early 2010s	~5–7 years
REMUS 6000	WHOI/Hydroid (USA)	Late 1990s	Early 2000s	Mid-2000s	~5–8 years
Echo Voyager / Orca XLUUV	Boeing/HII (USA)	~2012–2016	2017–2022	2023 (first delivery)	~7–11 years
Manta Ray	Northrop Grumman/DARPA (USA)	2019	2022–2024	2024–2025 (anticipated)	~5–6 years
Qilin ARV	CSSC (China)	2016	2019–2021	Late 2021	~5 years
Ghost Shark XL-AUV	Anduril Industries (Australia)	2022	2023–2024	Mid–Late 2025 (planned)	~3–4 years
Poseidon 6000	CSIC (China)	Early 2020s	2024–2025	Expected 2025	~3–5 years (estimate)

- **REMUS 6000**: A product of iterative research in academic and defence contexts, its development spanned from early experiments to a mature commercial product.
- **Echo Voyager/Orca XLUUV**: These platforms show a longer cycle due to the complexities of large, deep-diving unmanned systems intended for critical naval operations.
- **Manta Ray**: Represents a more recent effort focused on modularity and long-range, autonomous capability for military applications.
- **Qilin ARV**: A Chinese hybrid system combining autonomous and remotely controlled modes; its development was relatively rapid.
- **Ghost Shark**: An example of an ultra-fast-track development cycle aimed at fulfilling emerging defence requirements with a focus on stealth and modularity.
- **Poseidon 6000 AUV**: One of the latest entries in a series of Chinese AUVs designed for deep-sea exploration, with timelines still under refinement.

12.9 CONCLUSION

A variety of case studies of UUV development across the world have been discussed and their development journey has been outlined. The inputs about the timelines involved would offer users and manufacturers some appreciation of the technical and operational aspects of designing and deploying unmanned undersea vehicles.

REFERENCES

Blidberg, D. R. (2001). The Development of Autonomous Underwater Vehicles (AUV): A Brief Summary. *Proceedings of the IEEE ICRA*, *4*, 122–129.

Chuanren, C. (2025, September 4). China Rolls Out New Unmanned Underwater Systems in Parade. *Asian Military Review*. Retrieved September 14, 2025, from https://www.asianmilitaryreview.com/2025/09/china-rolls-out-new-unmanned-underwater-systems-in-parade-foc/

CSIR-NIO. (2023). *Autonomous Underwater Vehicles (AUV)*. CSIR-National Institute of Oceanography. Retrieved May 2024, from https://www.nio.res.in/research/technologies/autonomous-underwater-vehicles-auv

Dickinson, P. (2025, June 12). Ukraine Is Shaping The Future of Drone Warfare at Sea as Well as on Land. *Atlantic Council*. Retrieved September 2025, from https://www.atlanticcouncil.org/blogs/ukrainealert/ukraine-is-shaping-the-future-of-drone-warfare-at-sea-as-well-as-on-land/

Eckstein, M. (2023, November 9). US Navy Tests Sub-launched Drones While Industry Continues Designing. *Defense News*. Retrieved September 2025, from https://www.defensenews.com/naval/2023/11/08/us-navy-tests-sub-launched-drones-while-industry-continues-designing/

Elgar Desa, R. M. (2007). The Small AUV Maya: Initial Field Results. *International Ocean Systems, 11*(3), 1–9.

EMR Claight. (2025). *UUV Market Size and Share - Growth Analysis Report and Forecast Trends*. Expert Market Research Claight. Retrieved September 2025, from https://www.expertmarketresearch.com/reports/unmanned-underwater-vehicle-uuv-market

Exploring NUSTEM. (n.d.). *History of Unmanned Underwater Vehicles.* Exploring Naval Underwater STEM. Retrieved November 2025, from https://nustem.bridgeport.edu/history-of-uuv/

Fedasiuk, R. (2021, August 22). How China is Militarizing Autonomous Underwater Vehicle Technology. *The Maritime Executive.* Retrieved September 2025, from https://maritime-executive.com/editorials/how-china-is-militarizing-autonomous-underwater-vehicle-technology

Fotheringham, D. (2021, December). *The REMUS Legacy: Celebrating 20+ years of REMUS.* Marine Technology Reporter. Retrieved September 2025, from https://www.maritime-magazines.com/marine-technology/202112/the-remus-legacy/

GAO. (2022). *Extra Large Unmanned Undersea Vehicle: Navy Needs to Employ Better Management Practices to Ensure Swift Delivery to the Fleet.* US Government Accountability Office.

General Dynamics. (2015). *Bluefin-21.* General Dynamics Mission Systems. Retrieved April 2025, from https://epic.awi.de/id/eprint/40899/1/Bluefin-21-AUV.pdf

General Dynamics. (2024). *Bluefin-21 Unmanned Underwater Vehicle (UUV).* General Dynamics Mission Systems. Retrieved May 2024, from https://gdmissionsystems.com/products/underwater-vehicles/bluefin-21-autonomous-underwater-vehicle

GMI. (2024, May). *Autonomous Underwater Vehicle (AUV) Market Size.* Global Markets Insight. https://www.gminsights.com/industry-analysis/autonomous-underwater-vehicle-auv-market

Goldstein, L. J. (2021, March 7). UUVs are Giving China New Means of Naval Attack. *National Interest.* Retrieved May 2024, from https://nationalinterest.org/blog/reboot/uuvs-are-giving-china-new-means-naval-attack-179452

Hagen, O., Anonsen, K., & Mandt, M. (2010). The HUGIN Real-time Terrain Navigation System. In *OCEANS 2010* (pp. 1–7). Sydney: IEEE.

HII. (2025). *REMUS UUVs.* HII. Retrieved June 2025, from https://hii.com/what-we-do/capabilities/unmanned-systems/remus-uuvs/

Hofoss, E. (2021, November 12). *The Story of Hugin – An Autonomous Underwater Vehicle.* FFI - Norwegian Defence Research Establishment. Retrieved September 2025, from https://www.ffi.no/en/news/the-story-of-hugin--an-autonomous-underwater-vehicle

Hydroid Inc. (2012, July). *Marine Technology Reporter.* Retrieved June 2025, from www.seadiscovery.com: https://magazines.marinelink.com/Magazines/MarineTechnology/201207/page/56

iDEX. (2025). *iDEX Challenges.* Innovations for Defence Excellence. Retrieved May 2025, from https://idex.gov.in/challenges

Jalving, B., Gade, K., Hagen, O., & Vestgård, K. (2004). A Toolbox of Aiding Techniques for the HUGIN AUV Integrated Inertial Navigation System. *Modeling, Identification and Control*, 173–190.

Jalving, B., Vestgård, K., & Storkersen, N. (2002). Detailed Seabed Surveys with AUVs. In G. Griffiths (Ed.), *Technology and Applications of Autonomous Underwater Vehicles* (pp. 179–202). CRC Press.

Jianguo, W. (2018). Analysis of the Current Status and Problems of the Development of China's Autonomous Underwater Robots and Core Components. In *Oceanology International.* Hebei University of Technology.

Kongsberg. (2009, June). *Naval AUV product range The HUGIN & REMUS Family.* www.kongsberg.com: https://www.kimerius.com/app/download/5785529677/The+Hugin++Remus+family.pdf

Kongsberg. (2024, September). *Smashing AUV Records - HUGIN Endurance Completes a Multi-Week Fully Autonomous Mission.* Kongsberg. Retrieved May 2025, from https://www.kongsberg.com/newsroom/news-archive/2024/smashing-auv-records-hugin-endurance-completes-a-multi-week-fully-autonomous-mission/

Kongsberg. (2025). *Hugin AUV.* Kongsberg. Retrieved July 2025, from https://www.kongsberg.com/discovery/autonomous-and-uncrewed-solutions/auv/

Korulla, M. (2016, April). *Autonomous Underwater Vehicles: Roadmap for a Future Ready Naval Force.* International Seminar on 'Make in India' Paradigm - Roadmap for a Future Ready Naval Force. FICCI, New Delhi.

Kullashri, V. (2022, December 3). Unmanned Underwater Vehicle – The Invincible 'Varun Astra'. *Raksha-Anirveda.com.* Retrieved May 2025, from https://raksha-anirveda.com/unmanned-underwater-vehicle-the-invincible-varun-astra/

Lundquist, E. (2021, April 6). *ICEX 2020: U.S. Navy Deploys Autonomous Bluefin-21 UUV Under The Ice.* General Dynamics Mission Systems. Retrieved May 2025, from https://gdmissionsystems.com/articles/2021/04/26/in-the-news-us-navy-deploys-autonomous-uuvs-under-the-ice

Lundquist, E. (2023, Q2). *Bluefin Robotics Celebrates its 25th Anniversary.* Ocean Robotics Planet. Retrieved May 2025, from https://www.rovplanet.com/tportal_upload/md_publications/rovplanet_35.pdf

Manley, J. (2019, January). Unmanned Maritime Vehicles: 25 Years, Hundreds of Milestones. *Marine Technology News*, pp. 26–31.

Marine Technology News. (2024, February 9). Kongsberg Discovery Wins US Navy DIU Contract. *Marine Technology News.* Retrieved May 2025, from https://www.marinetechnologynews.com/news/kongsberg-discovery-contract-634492

Marine Technology News. (2024, September 4). *HUGIN Endurance AUV Smashes Records in Multi-week Fully Autonomous Mission.* Marine Technology. Retrieved May 2025, from https://www.marinetechnologynews.com/news/hugin-endurance-smashes-records-639764

Maurya, P., Morishita, H., & Pascoal, A. (2022). A Path-Following Controller for Marine Vehicles Using a Two-Scale Inner-Outer Loop Approach. *Sensors, 22*, 4293.

MoES. (2024). *Deep Ocean Mission.* Ministry of Earth Sciences, Govt of India. Retrieved May 2025, from https://moes.gov.in/schemes/dom?language_content_entity=en

Na, C. (2025, March 3). *China Launches Construction of Cold-seep Ecosystem Research Facility.* Chinese Academy of Sciences. Retrieved May 2025, from https://english.cas.cn/newsroom/cas_media/202503/t20250303_902833.shtml

Naval Technology. (2023, March 21). *REMUS 300 Unmanned Underwater Vehicle.* Naval Technology. Retrieved September 2025, from https://www.naval-technology.com/projects/remus-300-unmanned-underwater-vehicle-uuv/

Newman, P., Westwood, R., & Westwood, J. (2007, October 27). *Market Prospects for AUVs.* Hydro International. Retrieved July 2025, from https://www.hydro-international.com/content/article/market-prospects-for-auvs

NMICPS. (2023). *National Mission on Interdisciplinary Cyber-Physical Systems.* NMICPS, Dept of Science & Technology, Govt of India. Retrieved May 2025, from https://nmicps.gov.in/#scope

Oceaneering. (2025). *Hugin Superior 6000.* Oceaneering. Retrieved September 2025, from https://www.oceaneering.com/products/hugin-superior-6000/

O'Rourke, R. (2025). *Navy Large Unmanned Surface and Undersea Vehicles: Background and Issues for Congress.* Congressional Research Service.

Ray, A. (2025). *Underwater Robotics - Engineering and Applications.* New Delhi: National Maritime Foundation.

Ray, A., Singh, S. N., & Seshadri, V. (2011, June, 29–30). *Underwater Gliders - Force Multipliers for Naval Roles. Warship 2011: Naval Submarines and UUVs.* Bath, UK: Royal Institution of Naval Architects.

Sathianarayanan, D. (2013). *Deep Sea Qualification of Remotely Operable Vehicle* (ROSUB 6000). IEEE International Underwater Technology Symposium (UT). Tokyo, Japan, pp. 1–7.

SBIR-STTR. (2020, July 7). *SBIR-STTR-Success: Bluefin Robotics Corp.* Retrieved May 2024, from https://www.sbir.gov/success/sbir-sttr-success-bluefin-robotics-corp

Sea Grant. (n.d.). *AUV LAB – Odyssey Class History.* Sea Grant MIT. Retrieved September 2025, from https://seagrant.mit.edu/auv-odyssey-class/

Shome, S. E. (2010). Autonomous Underwater Vehicle for 150m Depth: Development Phases and Hurdles Faced. In P. E. Vadakkepat (Ed.), *Trends in Intelligent Robotics. FIRA 2010* (Vol. 103). Berlin: Springer.

Skopljak, N. (2022, July 13). Hugin AUVs bring over €43 million to Kongsberg. *Offshore-Energy.biz.* Retrieved May 2025, from https://www.offshore-energy.biz/hugin-auvs-bring-over-43-million-to-kongsberg/

Stommel, H. (1989, April). The Slocum Mission. *Oceanography.*

Strachan, D. R. (2019, November 22). China Enters the UUV Fray. *The Diplomat.* Retrieved October 2025, from https://thediplomat.com/2019/11/china-enters-the-uuv-fray/

Taylor, N. E. (2014, April 17). A Robot Dives into Search for Malaysian Airlines Flight. *MIT Spectrum.* Retrieved May 2025, from https://news.mit.edu/2014/robot-dives-search-malaysian-airlines-flight

TechSci Research. (2025, March). Asia-Pacific AUV and ROV Market. *TechSci Research.* https://www.techsciresearch.com/report/asia-pacific-auv-and-rov-market/27738.html

US Department of Defense. (2018). *Unmanned Systems Integrated Roadmap FY2017-2042.* Washington, DC: Office of the Under Secretary Of Defense (Acquisition Technology and Logistics).

Wennberg, A. (2023). *Taking A Deep Dive to Find Important Answers.* Kongsberg. Retrieved September 2025, from https://protect.kongsberg.com/article_taking-a-deep-dive-to-find-important-answers/

WHOI. (2025). *REMUS.* Woods Hole Oceanographic Institution. Retrieved August 2025, from https://www.whoi.edu/what-we-do/explore/underwater-vehicles/auvs/remus/

13 Future Prospects

Rapid technological change is the only constant in today's world, affecting every aspect of human life and impacting the planet as well. The significant role played by the oceans in maintaining the Earth's climatic equilibrium is becoming more evident as researchers look deeper and wider. Marine robotics and underwater robotics will increasingly be the instruments of choice of scientists, oceanographers, and ocean engineers in the incessant human quest for knowledge, survival, and resources.

This chapter summarises future areas of technological development in underwater robotics. This discussion could possibly result in mixing the 'desirable', the 'essential', and the 'possible'. However, that may be acceptable since these represent a range of capabilities that are not available uniformly across industry segments, or firms, or nations. What is routine for one may be an aspiration for another. In that spirit, the future is explored in terms of the technology areas that have been discussed in the preceding chapters, as well as operational underwater capabilities that could become the new normal in the years ahead.

13.1 EMERGING ROLES

The past decade has seen UUV technology explode in scope and ambition. Military, scientific, and commercial UUVs have now charted continental shelves, supported disaster recovery, and increased human reach for undersea applications without deploying divers or manned submersibles. In the Ukraine-Russia conflict, both sides have used a mix of maritime drones for surveillance, sabotage, and precision strikes against ships and infrastructure, accelerating global development of counter-UUV technologies. A few emerging areas of application are listed here.

- **Inland Inspections**: UUVs are being deployed increasingly in inland water bodies for routine underwater infrastructure inspection, resource mapping, and ecosystem health monitoring. Such tasks were earlier cost-prohibitive or undertaken manually.
- **Cross-Domain Solutions**: Use cases for 'cross-domain' operations have spurred solutions for trans-medium UUVs that can transition from underwater to aerial operation and vice versa.
- **Enhanced Autonomy and Decision Support**: Modern AUVs blend classic pre-planned operations with the flexibility to improvise, adapt, and collaborate with other manned and unmanned systems. AI techniques could progressively enable full mission autonomy.
- **Swarming Algorithms and Distributed Intelligence**: Dynamic swarms of small UUVs could act in concert, using distributed algorithms to cover vast operating areas for surveillance, mapping, or intra-swarm warfare.

DOI: 10.1201/9781003737476-13

These fleets offer resilience, and the loss of some could be compensated by the rest of the swarm.
- **Weaponisation and Direct Strike Roles**: The US, Russia, Iran, North Korea, and Ukraine have fielded or tested various armed UUVs, including explosive-laden kamikaze drones, blurring lines between surveillance and strike.
- **Networks and Data Integrity**: As underwater 'cloud' networks emerge, ensuring the integrity and resilience of UUV data links is paramount. Robust, standardised, hardened, and encrypted protocols are required for cross-domain data exchange and coordination.
- **Loitering Munitions**: This concept is borrowed from aerial drones and may find undersea analogues capable of station-keeping for days, monitoring targets, and, when necessary, acting autonomously to neutralise high-value threats.

13.2 EMERGING SENSOR TECHNOLOGY

As AUV missions become more demanding and the need for real-time, high-resolution underwater data increases, sensor technology is evolving rapidly. Underwater sensor technology is moving towards systems that are smaller, smarter, and more energy-efficient.

13.2.1 Miniaturised Multi-parameter Sensor Systems

Developments in MEMS, microfluidics and complementary metal–oxide–semiconductor (CMOS) technologies are enabling the production of integrated multi-parameter sensors that can measure several environmental variables simultaneously (Carder et al., 2001). These 'lab-on-a-chip' devices can combine:

- **Temperature, conductivity, and depth measurements** (CTD)
- **Chemical sensing** (e.g., pH, dissolved oxygen, nitrates, and heavy metals)
- **Biological parameters** (e.g., chlorophyll fluorescence for phytoplankton)

These systems not only reduce size and weight but also simplify the integration process by providing a single, unified sensor package. New materials and manufacturing techniques allow for ultra-compact sensor housings that are robust against pressure and corrosion. For instance, composite materials and advanced polymer coatings are being used to protect delicate sensor elements while minimising weight and size.

13.2.2 Smart and Adaptive Sensors

Future sensors are expected to incorporate on-board processing and machine-learning algorithms, with advanced features (Antonelli, 2018).

- **Self-calibration**: Sensors that automatically calibrate themselves to account for drift or environmental changes. Advanced optical sensors now incorporate algorithms that adjust for lens distortion, chromatic aberration, and backscatter in real time, improving data reliability over long-duration missions.
- **Adaptive Sampling**: Sensors that dynamically adjust their sampling rates based on real-time environmental variability or mission-critical events. This adaptive behaviour allows an AUV to conserve energy during periods of low variability and increase data acquisition during events such as phytoplankton blooms or pollutant dispersal.
- **Embedded AI**: Integration of AI to perform preliminary data analysis on board, reducing the amount of data that must be transmitted and allowing for autonomous decision-making. By incorporating artificial intelligence (AI) algorithms into sensor systems, AUVs can perform real-time data analysis and make autonomous decisions about sensor activation, data compression, and anomaly detection. For example, machine-learning-based image enhancement techniques can automatically correct for underwater colour distortion and blur.

13.2.3 Advanced Materials and Transducers

Innovations in materials science are expected to yield improvements in transducer performance.

- **Nanomaterials and Metamaterials**: These materials may offer enhanced acoustic transmission and reception properties, leading to improved sensitivity and bandwidth for acoustic modems.
- **Miniaturised Optical Components**: Advances in laser diodes, photodetectors, and beam steering technologies will improve the range and reliability of optical communication systems.
- **Robust, Low-Power Electronics**: The development of low-power, high-efficiency electronic components will further reduce the energy footprint of underwater communication devices, enabling longer mission durations.

13.2.4 Advanced Optical Imaging

Optical sensor technology is rapidly advancing in areas such as high-resolution and hyperspectral imaging, and attempts to use LiDAR underwater (Rice & Green, 2008).

- **High-Resolution Imaging**: New CMOS sensors with enhanced low-light performance and higher dynamic range allow for better image quality even in the murky underwater environment. Synthetic-Aperture Sonar (SAS) systems continue to evolve, providing finer resolution and improved image quality for seabed mapping, mine detection, and infrastructure inspection.

The integration of SAS with AUV platforms is enabling detailed 3D reconstructions of underwater environments (Wadoo & Kachroo, 2011).

- **Underwater Hyperspectral Imaging**: Optical sensors that capture a wide spectrum of light wavelengths are emerging, providing detailed spectral signatures that can be used for environmental monitoring, habitat classification, and even chemical detection. They can provide detailed information about water composition, biological activity, and the mineralogy of the seafloor.
- **Underwater LiDAR**: Although challenging because of scattering effects, advances in laser technology and signal processing are expected to make underwater LiDAR increasingly feasible for 3D mapping applications in clear water conditions (Roberts & Sutton, 2013).

13.2.5 Hybrid Acoustic-Optical Sensors

Hybrid sensors can offer complementary data—acoustic sensors provide robust range and imaging in turbid waters, while optical sensors deliver high-resolution imagery in clearer conditions (Balasuriya & Ura, 2002). To overcome the limitations of individual modalities, researchers are developing hybrid sensors that integrate both acoustic and optical components.

- **Dual-Mode Imaging**: Such sensors can operate in both optical and acoustic modes, switching between them depending on water clarity and ambient light conditions (Zhang et al., 2022). In turbid water, acoustic imaging provides reliable data; in clear water, optical imaging offers high resolution (Yuan et al., 2017; Zhang et al., 2022).
- **Sensor Fusion**: Combining data from acoustic and optical sensors can yield a more complete understanding of the environment. Advanced sensor fusion algorithms are being developed to integrate disparate data types into coherent models of the seafloor and water column (Cecchi et al., 2005).
- **Extensible Payload Modules ('Whiskers')**: These modules are designed for easy integration into AUVs, offering the flexibility to combine optical, acoustic, and chemical sensors into a single payload package. Such modules emphasise rapid reconfiguration and adaptive mission planning.

13.2.6 Biomimetic and Bio-inspired Sensors

Taking inspiration from marine organisms, researchers are developing sensors that mimic biological systems (Hasan, 2017).

- **Biomimetic Optical Sensors**: Inspired by the eyes of deep-sea creatures, these sensors aim to achieve superior low-light performance and wide dynamic ranges.

- **Acoustic Mimicry**: Sensors that mimic the hearing mechanisms of marine mammals could lead to improved detection capabilities in noisy environments.
- **Geophysical Sensors**: Sensors for gravimetry or magnetometry may become viable in the future (Teixeira & Pascoal, 2007).
- **Self-healing Materials**: Emerging research into self-healing polymers and coatings could extend the lifespan of sensor housings by automatically repairing minor damage caused by pressure or biofouling. This could extend sensor lifespan and reduce maintenance.

13.3 EMERGING COMMUNICATION TECHNOLOGY

There have been significant advancements in underwater communication technologies, driven by interdisciplinary research and the need for higher data throughput, longer ranges, and improved reliability (Defense Science Board, 2015). Emerging hybrid solutions and developments in materials, signal processing, and artificial intelligence (AI) are advancing the limits of underwater communication.

13.3.1 Acoustic Communication

- **High-Frequency Acoustic Modems**: New acoustic modems have been developed that offer higher bandwidths (up to tens of kilobits per second) over moderate ranges (up to a few kilometres) (Hasan, 2017). These devices benefit from improved modulation techniques such as OFDM and advanced error correction algorithms.
- **Low-Power Acoustic Networks**: Research into low-power acoustic communication has enabled the development of sensor networks for environmental monitoring. These networks can operate for extended periods on battery power, transmitting data intermittently (Randeni et al., 2022).
- **Adaptive Acoustic Systems**: Advances in signal processing have allowed acoustic systems to adapt to changing environmental conditions, dynamically adjusting frequency, power, and modulation schemes to optimise performance (Kim, Lee, & Myung, 2020).

13.3.2 Optical Communication

- **Blue/Green Laser Systems**: Modern optical communication systems use blue and green lasers to take advantage of the lower attenuation at these wavelengths in seawater. These systems can achieve data rates in the megabit-per-second range over distances of tens of metres in clear water.
- **Optical Wireless Communication (OWC)**: OWC systems are being developed for short-range high-bandwidth communication between underwater vehicles and between vehicles and surface stations. Recent prototypes have demonstrated reliable communication in controlled conditions (Antonelli, 2018).

- **Alignment and Beam Steering Technologies**: Innovations in beam steering and automatic alignment systems help maintain optical links even in the dynamic underwater environment. These systems use advanced sensors and feedback control to continuously adjust the laser beam direction.

13.3.3 Electromagnetic Communication

- **Extremely Low Frequency (ELF) Developments**: Research continues into improving ELF communication systems to enable secure communication with submerged assets over long ranges. While these systems remain limited in bandwidth, they are critical for strategic communication in military applications.
- **Integration with Other Modalities**: EM communication is increasingly being considered as part of a hybrid system, where it can provide backup or supplementary communication when acoustic channels are disrupted.

13.3.4 Hybrid and Multimodal Systems

Hybrid communication systems aim to leverage the strengths of multiple modalities to overcome individual limitations.

- **Adaptive Switching**: Some systems are designed to switch between acoustic and optical communication modes based on environmental conditions and data rate requirements. For example, an AUV might use optical communication when in close proximity to a base station to transfer high-bandwidth data and revert to acoustic communication for long-range transmission.
- **Adaptive Multimodal Platforms**: Emerging platforms are being designed to dynamically switch between acoustic, optical, and EM channels. These systems use AI-driven decision algorithms to select the optimal channel based on current environmental conditions, operational range, and data rate requirements. Advanced digital signal processing will enable real-time, adaptive switching between communication modes based on environmental factors and mission requirements (Ghabcheloo et al., 2009).
- **Integrated Systems**: By integrating multiple communication modalities, advanced autonomous systems can ensure robust and reliable data transfer under varying conditions. AI-driven algorithms can help select the best communication channel based on real-time analysis of environmental factors, required data rate, and range (Inzartsev & Pavi, 2009). Development of integrated transceiver units capable of operating across multiple frequency bands will reduce system complexity and power requirements.

13.4 COLLABORATIVE OPERATIONS

While the potential of collaborative AUV operations is substantial, several challenges and research opportunities remain. These include limitations in underwater communication, robust autonomy, energy constraints, and standardisation of interoperability and data exchange protocols.

13.4.1 Advanced Swarm Intelligence

Future AUV swarms are expected to incorporate more sophisticated machine learning techniques and swarm intelligence algorithms. These systems could learn from past missions, optimise cooperation strategies in real time, and even predict environmental changes based on historical data (Figure 13.1). Enhanced autonomy will enable swarms to operate in increasingly complex and unstructured environments (Qiu et al., 2020).

13.4.2 Improved Sensing and Data Fusion

Ongoing research in sensor technology—ranging from improved sonar arrays to novel optical imaging systems—will further enhance the capabilities of AUV swarms. Advanced data fusion techniques that combine information from diverse sensor modalities will enable a more complete understanding of the underwater environment, thereby enhancing mission decision-making.

FIGURE 13.1 Demonstration of AUV swarm by NSTL, India.

Source: DRDO.
Notes: https://www.pib.gov.in/PressReleasePage.aspx?PRID=2190096, (PIB, 2025).

13.4.3 Hybrid and Heterogeneous Swarms

The future of collaborative operations likely lies in heterogeneous swarms, where different types of underwater vehicles (e.g., AUVs, gliders, and even unmanned surface vehicles) work together to achieve a common mission. Such hybrid systems can capitalise on the strengths of each platform—for example, AUVs for high-resolution data collection and gliders for long-endurance, low-power surveying. Research into the coordination of heterogeneous systems is an active and promising field.

13.4.4 Surface and Satellite Systems

To overcome communication and navigation challenges, future AUV swarms may increasingly integrate with surface platforms and satellite networks. These integrated systems can provide a relay for data, assist in global positioning (e.g., through GPS-enabled surface buoys), and enable remote real-time control. European initiatives are already exploring such integrated architectures, aiming to develop a seamless continuum from underwater to space-based systems.

13.4.5 Autonomous Mission Reconfiguration

One of the most exciting prospects for the future is the development of swarms that can autonomously reconfigure their mission plans in response to real-time data. For example, if a subset of the swarm detects an anomaly or an area of interest (such as a pollutant plume or an underwater hazard), the swarm could automatically allocate more resources to investigate that area while maintaining overall mission balance. This level of adaptive mission management would significantly enhance the utility and efficiency of underwater operations.

13.4.6 Standardisation and Interoperability

As underwater communication systems evolve, there is a growing need for standardised protocols and interfaces. International bodies and industry consortia are working to establish common standards for underwater data communication (e.g., IEEE and ITU recommendations for underwater acoustic communication).

Standardised interfaces will enable unmanned systems from different manufacturers to communicate seamlessly, facilitating multi-vendor, multi-domain network-centric operations (Griffiths, 2003).

Enhanced certification processes will be required to ensure that emerging technologies meet safety, reliability, and performance criteria in the challenging underwater environment.

13.4.7 Internet of Underwater Things (IoUT)

Underwater autonomous systems of the future would increasingly involve networked sensor and vehicle systems, commonly referred to as the Internet of Underwater

Things (IoUT) or Subsea Internet of Things (SIoT) (Nkenyereye, Nkenyereye, & Ndibanje, 2024). Such underwater networks would not be able to use electromagnetic waves, and without physical connections, would need to rely on acoustic communication, making them a futuristic proposition. Underwater communication reliability, range and fidelity are crucial for the viability of such networks (Gupta & Singh, 2017; Hyland, 2016).

Distributed sensor networks (such as 'PLUSNet' and 'Seaweb' (Rice & Green, 2008) of the US ONR/ DARPA) use concepts that are now included under IoUT/ Underwater Wireless Sensor Networks (UWSNs). They leverage advances in communication and edge computing (Nkenyereye, Nkenyereye, & Ndibanje, 2024).

Current areas of development in the field of UWSNs/SIoT/IoUT are briefly mentioned below.

- **Interconnected/Cloud-Connected Sensors**: Future sensor networks will be fully integrated, with sensors communicating with each other and with cloud-based systems via acoustic or optical modems. Advances in acoustic and optical communications now allow underwater sensor networks to link with onshore cloud platforms for real-time data aggregation and analysis. This connectivity enhances operational decision-making and enables large-scale environmental monitoring (Qiu et al., 2020; Heshmat et al., 2025).
- **Edge and Fog Computing**: With limited underwater communication bandwidth, processing data at the edge (on the sensor or AUV itself) reduces the need for high-bandwidth transmission. Emerging edge computing platforms are designed for low power and high efficiency in underwater applications (Zuluaga et al., 2022).
- **Interoperability and Standardised Protocols**: Efforts are underway to develop standardised communication and data formats for underwater sensor networks, which will improve interoperability and scalability. Standardised interfaces and modular payload designs simplify sensor selection and integration across diverse AUV platforms.

13.4.8 Integration of AI/ML

AI and machine learning are increasingly applied to optimise underwater communication, network management, and networked operations.

- **AI for Communication/Channel Optimisation**: Machine learning algorithms process sensor data to predict channel conditions, adjust transmission parameters in real time, and manage network traffic among multiple autonomous vehicles. Such algorithms can predict channel conditions (such as ambient noise, multipath interference, and water clarity) and dynamically adjust transmission parameters (Nichols et al., 2020).
- **Network Management**: Autonomous systems can use AI to manage multimodal communication networks, deciding in real time which modality to use for data transmission to ensure reliability and energy efficiency.

- **Network-Centric Operations**: Future naval and industrial autonomous systems are expected to operate as part of integrated, network-centric architectures. Advances in underwater communication protocols and middleware (such as DDS and ROS) facilitate the coordination and data sharing necessary for these distributed systems.
- **Data Compression and Error Correction**: AI-driven adaptive algorithms can improve data throughput by optimising compression schemes and robust error correction tailored to underwater conditions.
- **Energy Efficiency**: Low-power design and energy harvesting techniques are being integrated with communication systems to extend operational endurance, particularly in long-term environmental monitoring networks.

13.5 LAUNCH AND RECOVERY

The future of LARS is likely to be defined by increased automation, with AI and machine learning integrated into control systems to improve the system's responsiveness to dynamic environmental conditions. For example, adaptive control algorithms can learn from sensor inputs and adjust winch speeds or telescopic arm positioning in real time. As these technologies mature, we can expect LARS to operate with minimal human intervention, thereby increasing safety and operational efficiency.

The integration between USVs, UUVs and their LARS is expected to become tighter. USVs with on-board LARS capabilities can serve as mobile recovery platforms that can launch and retrieve AUVs autonomously. Such integrated systems not only reduce personnel risk but also allow for operations in more challenging conditions where manned vessels may be at risk.

13.6 UNDERWATER NAVIGATION

While significant progress has been made in reducing drift and enhancing navigation accuracy through sensor fusion, several challenges remain. Environmental variability, such as changing currents, temperature gradients, and acoustic multipath effects, can still introduce errors. Advanced fusion algorithms (e.g., particle filters and graph-based SLAM) are computationally intensive. Techniques for on-the-fly calibration and self-diagnosis of sensor biases will further improve navigation accuracy. Emerging sensor technologies, such as quantum magnetometers or low-power optical sensors, hold promise for further reducing drift. Their integration into existing sensor fusion frameworks will be an important area of future research.

Some focus areas for future research on underwater navigation are suggested below (Leonard & Bahr, 2016).

- **Adaptive and Robust Filtering**: Developing algorithms that can dynamically adjust to environmental changes and sensor degradation.
- **Integration of Emerging Sensors**: Incorporating quantum sensors, low-power optical systems, and advanced magnetometers to further reduce drift.

- **Computational Efficiency**: Leveraging hardware acceleration (e.g., GPUs, FPGAs) and distributed processing for real-time fusion in resource-constrained environments.
- **Cooperative Navigation**: Expanding multi-AUV cooperative sensor fusion to share data and correct drift collaboratively, especially in large-scale missions (Bahr, Leonard, & Fallon, 2009).
- **Long-Term Autonomy**: Focusing on systems that maintain high accuracy over days or weeks, crucial for deep-sea exploration and environmental monitoring.
- **Enhanced Outlier Rejection**: Refining methods for detecting and discarding spurious sensor measurements to prevent degradation of the navigation solution.
- **Collaborative Navigation**: In multi-AUV systems, inter-vehicle communication and cooperative sensor fusion can provide redundancy and improve overall system robustness. Developing standardised protocols and decentralised fusion algorithms will be key to leveraging the full potential of collaborative underwater navigation.

13.7 EMERGING SOFTWARE ARCHITECTURE

As AUVs become more sophisticated and their operational roles expand, several emerging trends and future developments are poised to transform their design and software architecture. As AUVs evolve to tackle more complex tasks, emerging trends in middleware, data protocols, and cognitive control architectures will further improve their performance and reliability.

Future efforts should focus on developing updated standards that address the full spectrum of challenges posed by modern underwater vehicles. This includes creating new guidelines for autonomous software and AI-driven control systems, integrating cybersecurity into the certification process, and establishing internationally harmonised standards that facilitate interoperability and global collaboration.

Such advances will not only improve the safety and reliability of underwater vehicles but will also accelerate innovation and reduce the costs and risks associated with developing and deploying these complex systems.

13.7.1 Enhanced Modularity

Future AUVs are expected to feature even greater modularity, both in hardware and software.

- **Hardware Upgradability**: Development of universal modules that adhere to strict standardisation guidelines will enable rapid replacement and upgrade of components without redesigning the entire system.
- **Software Microservices**: Microservices architectures that are common in cloud computing could be extended to AUVs. Here, each functional

module (e.g., navigation, sensor fusion, mission planning) can be developed, deployed, and updated independently.
- **Middleware Evolution**: The next generation of middleware will further abstract hardware differences, providing seamless integration across heterogeneous platforms. Open-source initiatives and standard bodies are expected to drive these improvements.

13.7.2 Advanced Backseat Driver and Autonomy

With increasing computational power and advances in artificial intelligence, the backseat driver role will evolve significantly:

- **Cognitive Systems**: Future supervisory modules will leverage machine learning to adapt mission plans based on real-time environmental data, historical performance, and predictive models.
- **Context Awareness**: Enhanced sensor fusion and context-aware systems will allow the backseat driver to better understand its environment and make more informed decisions.
- **Collaborative Autonomy**: In missions involving multiple AUVs, backseat drivers on individual vehicles may communicate and coordinate via distributed algorithms to optimise overall mission performance.
- **Self-Healing Architectures**: The supervisory layer may incorporate advanced fault diagnosis and recovery techniques that allow the system to reconfigure itself automatically in response to failures.

13.7.3 Standardisation and Interoperability

Standardisation efforts are ongoing across the robotics and maritime industries (Costanzi, Fenucci, & Manzari, 2020). Future AUV systems will benefit from:

- **Global Interface Standards**: Adoption of common standards (such as those developed by AUTOSAR for automotive systems or ROS for robotics) in underwater applications will streamline integration between modules from different vendors.
- **Interoperable Data Protocols**: Standard data exchange formats and communication protocols will enable AUVs to interact with external networks, cloud services, and other vehicles more seamlessly.
- **Certification and Safety Standards**: As AUVs are deployed in critical applications, there will be an increasing focus on functional safety (e.g., ISO 26262 for automotive systems adapted for maritime use) and certification frameworks that ensure reliability and fault tolerance.

13.7.4 Fault Tolerance and Resilience

Fault tolerance will continue to be a major focus as AUVs are expected to operate in environments where maintenance or retrieval is difficult:

- **Redundant Architectures**: Future systems may incorporate not only hardware redundancy but also software-level redundancy. Virtualisation and containerisation techniques can isolate faults and allow a backup instance to take over.
- **Predictive Maintenance**: Using continuous diagnostics and machine learning models, AUVs can predict failures before they occur, scheduling maintenance or reconfiguring systems to avoid mission failure.
- **Distributed Control**: Instead of relying on a single mission computer, future designs may adopt a distributed control architecture in which multiple processors share the load and provide mutual backup.

13.7.5 Mission Computer

The mission computer is central to an AUV's operations, and future developments will likely include the following:

- **High-Performance Embedded Platforms**: Advances in low-power, high-performance processors and FPGAs will enable more complex algorithms to be executed on board without significant energy penalties.
- **Software-Defined Infrastructure:** Similar to concepts in network function virtualisation (NFV) and software-defined networking (SDN), future mission computers might support dynamic reconfiguration of software modules to better match changing mission requirements.
- **Edge and Cloud Integration**: Seamless integration between on-board processing and remote cloud computing resources will allow for more sophisticated data analysis and decision-making without overburdening the local hardware.
- **Interchangeable Modules**: By designing the mission computer as a series of swappable modules, updates and replacements can be made with minimal disruption to the overall system.

13.8 FUTURISTIC ENERGY SOLUTIONS

The future of underwater energy management may involve low-energy operations, energy harvesting from seawater, underwater recharging from seabed sources, and networked sensor and vehicle systems. Integration of AUVs with a Subsea Internet of Things may allow for centralised energy management, remote monitoring, and even coordinated recharging strategies across a fleet of vehicles. On-board processors can optimise energy usage by processing sensor data locally and only transmitting essential information, reducing communication energy costs.

13.8.1 Energy Harvesting Technologies

As energy remains a limiting factor for long-duration AUV missions, emerging sensor designs focus on reducing power consumption. Power efficiency remains a top priority for underwater sensors, and emerging technologies focus on reducing energy consumption and even harvesting ambient energy (Finn & Scheding, 2010).

New semiconductor technologies are pushing power requirements down into the microwatt range for some sensor functions.

Techniques such as piezoelectric energy harvesting, microbial fuel cells, and even miniature underwater turbines are being investigated to supplement battery power. Solutions are being developed to generate electricity from water flow or vibrations, or harness solar energy during surface intervals. These innovations can extend AUV mission durations and reduce the reliance on heavy battery packs (Hwangbo, Jeon, & Park, 2013).

13.8.1.1 Wave and Current Harvesters

Researchers in academic laboratories have developed small-scale prototypes of wave energy converters and underwater turbines (Yang An, 2025). These experiments, often conducted in wave tanks or flow channels, have demonstrated proof-of-concept systems capable of converting mechanical energy from water movement into electrical energy with promising efficiencies (Zeng & Wang, 2023).

13.8.1.2 Thermoelectric and Salinity Gradient Devices

Thermoelectric generators and reverse electro-dialysis systems have also been prototyped in the laboratory. Advances in nanostructured materials have led to TEGs with improved figures of merit, while novel membrane technologies are addressing the long-standing issues of fouling and energy loss in salinity gradient devices. Such experiments have provided critical insights into material performance and system scalability.

13.8.1.3 Field Deployments

Innovative research has integrated energy harvesters into mobile platforms such as AUVs. In one notable demonstration, an AUV equipped with piezoelectric energy harvesters was able to recharge its on-board battery system while in transit, effectively extending its operational range. This kind of hybrid approach, combining conventional battery storage with continuous energy harvesting, has the potential to revolutionise long-duration underwater missions. Several pilot projects have deployed fixed underwater energy harvesters in coastal areas to power sensor networks. These installations often combine multiple energy harvesting methods—such as a combination of tidal current turbines and thermoelectric generators—to provide continuous power under variable environmental conditions (Khan et al., 2017).

13.8.2 Underwater Recharging

Underwater recharging technologies aim to allow AUVs to replenish their energy supplies without the need to return to a base or surface vessel. Techniques such as magnetic induction or acoustic energy transfer are being explored for recharging AUVs while deployed. These systems are in the early research phase but hold promise for reducing downtime by allowing in situ recharging (Wang et al., 2024).

Research is underway to use acoustic or electromagnetic fields for wireless power transfer to sensors, potentially enabling in situ recharging of deployed sensor networks (Liu et al., 2024). Combining traditional batteries with energy-harvesting modules and in situ recharging infrastructure could offer a pathway to continuous or near-continuous operation.

Recent workshops and conferences have highlighted the importance of interdisciplinary approaches combining expertise in materials science, marine engineering, and power electronics to address the complex challenges of underwater energy harvesting.

Although still in early stages, such systems have the potential to reduce downtime and maintenance requirements.

13.8.2.1 Wireless Power Transfer (WPT)

Emerging work on underwater wireless power transfer (WPT) systems uses magnetic induction or acoustic energy transfer to recharge sensor nodes or AUVs (Mohsan et al., 2022; Chaudhary et al., 2023).

- **Magnetic Induction**: This method uses coils to transfer power via a magnetic field. Although efficient over very short distances, research is ongoing to increase the effective range (Wang et al., 2024).
- **Acoustic Power Transfer**: This approach uses sound waves to transmit energy. This method is still in its infancy, but could offer a means to recharge AUVs in situ.

13.8.2.2 Underwater Docking Stations

- **Mechanical Docking**: AUVs can be designed to autonomously dock with an underwater charging station. These stations can provide power via direct electrical contacts or through inductive coupling (Liu et al., 2024).
- **Automated Battery Swapping**: Some futuristic designs propose modular battery packs that can be automatically exchanged by an on-board mechanism during a docking procedure.

13.8.3 Microbial Fuel Cells (MFCs)

Microbial fuel cells use bacteria to catalyse the oxidation of organic matter in marine sediments, generating electrical energy as a by-product. This method leverages naturally occurring biochemical reactions to provide continuous, low-power energy suitable for powering remote sensors and communication nodes. They utilise the metabolic processes of bacteria to generate electricity from organic matter in seawater. Although currently limited by low power output, they offer an intriguing possibility for long-term, low-power sensor applications.

Recent improvements in MFC technology include the development of novel nanomaterials and conductive polymers to improve electron transfer efficiency. Modern MFC designs optimise the configuration of anodes and cathodes to maximise energy extraction while minimising biofouling. Pilot projects in coastal and estuarine environments have demonstrated the feasibility of MFCs for long-term, low-power applications (Ali, 2025).

However, MFCs typically produce very low power outputs, which limits their use to powering small electronic devices. Changes in sediment composition and organic content can affect performance, necessitating adaptive designs.

Field trials of microbial fuel cells in sediment-rich coastal zones have provided encouraging results. While the power outputs remain low, these systems have proven capable of powering low-power sensor nodes for extended periods. The low-maintenance nature of MFCs makes them particularly attractive for remote monitoring applications in environmentally sensitive areas.

13.9 CONTRASTING CLAIMS WITH REALITY CHECKS

The promise of 'breakthrough' underwater technologies is undeniably attractive, yet a careful analysis reveals that many of the claims are based on idealised assumptions or laboratory conditions that do not always translate to the field. A summary comparison of claims versus current performance for each technology is given in Table 13.1.

In each case, the gap between 'breakthrough' claims and field performance underscores the challenges inherent in deploying advanced technologies in harsh underwater environments. It is important to recognise that many of these technologies are still in active development. Ongoing research, improved integration techniques, and incremental advances in sensor materials and algorithms are steadily closing the gap between promise and reality.

Many of the technologies reviewed—particularly quantum navigation and high-speed optical communication—are still at relatively low technology readiness levels (TRLs). Laboratory demonstrations frequently showcase remarkable performance under ideal conditions, but scaling these systems for robust field use remains a work in progress.

Perhaps the greatest challenges lie not in the individual components but in their integration into comprehensive systems. Underwater operations demand that sensors, communication links, and navigation aids work seamlessly in a highly unpredictable and often hostile environment. The incremental improvements offered by each 'breakthrough' technology must be balanced against the increased system complexity and potential points of failure.

TABLE 13.1
Technology Claims and Current Reality

Technology	Key Claims	Current Reality
Optical/laser communication	Gigabit-per-second data rates, long-range (100 m+) in clear water, precise and secure links.	High data rates are achieved only under ideal conditions; typical ranges in coastal/turbid water are 10–20 m; alignment and environmental sensitivity limit robust deployment.
Acoustic vector sensors	Superior directional resolution, enhanced DOA estimation, improved target localisation, and beamforming.	Lab performance shows improved resolution; field trials reveal calibration challenges and modest improvements over conventional arrays; increased complexity and cost.
Quantum navigation	Ultra-low drift rates, highly precise inertial measurements independent of GPS, transformative for long-duration missions.	Lab prototypes demonstrate promising performance; field implementations remain experimental and sensitive to environmental disturbances; integration challenges persist.
Geomagnetic navigation	Passive, low-cost, globally available navigation with high accuracy using Earth's magnetic field.	Widely used as a complementary sensor; improvements over INS are modest; subject to local anomalies and interference; not sufficient as a standalone solution.

13.10 DEVELOPING AN ECOSYSTEM

Applications in the marine domain have less visibility than the multiplicity of aerial drones over land. The small size of the market leads to few takers in industry, and hence the negative cycle of low demand for indigenous specialised products. Articulation of 'blue-sky' needs is typically not a forte of the User. Demonstration of successful products is usually the best way to attract users, who can then envision further applications, which, in turn, drive production and technology development.

The measures proposed for developing national expertise in marine robotics would include the following:

- Gaining and developing expertise (knowledge and skills).
- Encourage R&D through funding.
- Utilisation of robotic solutions by users, boosting demand.
- Trials: Facilities, competence and experience.
- Scaling up components and solutions.
- Standards for production, evaluation and interoperability.
- Facilities for development and testing.

Therefore, there is a need to synergise efforts so that these required components of institutions, knowledge, skills, facilities, and a healthy market can be developed for marine robotics.

Building a robust ecosystem requires targeted actions spanning infrastructure, funding, user engagement, and policy support. Some suggestions are given below:

- The creation of a virtual or distributed Centre of Excellence (CoE) for marine robotics would catalyse coordinated research, facilitate knowledge sharing, and streamline integration efforts.
- Shipyards and industry should actively sponsor and invest in R&D aligned with operational needs and share waterfront testing facilities to expedite prototype validation. Academics must intensify collaborations via joint projects, internships, and research dissemination to ensure a skilled workforce pipeline. Institutional support also needs to extend to workforce training and skill development for operating sophisticated underwater systems.
- The industry must innovate aggressively, expanding component development for global markets and investing in new partnership models to share risk and resources. Innovation hubs and incubation centres with funding for dual-use technologies would sustain rapid advancements.
- Designated sea testing areas in biologically interesting locations, such as the islands of Lakshadweep, would encourage multidisciplinary field research in Indian academic institutions dealing with marine sciences.
- Comprehensive shore integration and testing infrastructure is essential for continual technology evaluation and product refinement. Trial areas can range from swimming pools to lakes, rivers, shallow harbours, and open sea. Conducting trials is also linked to support facilities such as launch and recovery systems, mother ships, measuring instruments, calibration facilities, etc. Such integration and testing facilities for uncrewed underwater vehicles are distributed across shipyards, private industry, government R&D labs, and defence forces, and these need to be shared and made available for common national goals.

13.11 LOOKING AHEAD

A futuristic article written in 1989 imagined a scenario of the 2020s, in which 'Underwater Gliders' would be deployed in dozens across the world's oceans (Stommel, 1989). They would collect bathymetric data for months at a time, periodically uploading via satellites. Such a comprehensive picture could be converted to reality today through processing centres and nodes ashore. Scientists, researchers, and industry users could work with data collected in real time, across oceans, without the forbidding cost and effort of putting to sea in a research ship to collect a few snapshots of data from the deep. Most of that vision has already come true.

The story of unmanned undersea vehicles is one of relentless innovation, spurred by changing operational demands and exponential advances in robotics, AI, and marine engineering. From hand-built prototypes charting Arctic seas to modern swarming fleets and strategic strike weapons, UUVs are reshaping the undersea battlespace and the global maritime landscape, promising a future where human presence is increasingly optional beneath the waves.

While current standards provide a strong foundation for underwater vehicle design and operation, the future of unmanned and autonomous underwater systems will require a greater role for certification agencies and more prescriptive, yet agile, design rules. Enhanced standards and certification processes will help ensure that as technologies evolve, vehicles operating beneath the waves remain safe, reliable, and capable of meeting the increasingly demanding challenges of both scientific exploration and industrial application (Aguzzi et al., 2022).

Today's youthful human resources, along with growing interest in robotics and the opportunities of the oceans, together offer tremendous opportunities in marine robotics. Unmanned and autonomous systems require the development of innovative vehicles and components, ranging from as small as a few kilograms, with applications not only in the oceans but also in any water body on land. In the years ahead, mankind will increasingly look to the oceans for food, energy and commerce. Unmanned and autonomous systems will play a significant role in these endeavours.

The Indian ecosystem for subsea robotics is not just a necessity but an imperative for strategic indigenous capability. The larger global market for marine unmanned systems also offers export opportunities to the Indian industry. To meet these aspirations, all stakeholders and agencies need to synergise efforts and collaborate to build up a robust and self-sustaining ecosystem.

REFERENCES

Aguzzi, J., Flogel, S., Marini, S., & Thomson, L. (2022). Developing Technological Synergies between Deep-sea and Space Research. *Elementa Science of the Anthropocene*, *10*(1), 1–19.

Ali, H. H.-N. (2025, January). Harnessing the Power of Microbial Fuel Cells as Pioneering Green Technology: Advancing Sustainable Energy and Wastewater Treatment Through Innovative Nanotechnology. *Bioprocess and Biosystems Engineering*, *48*, 343–366.

Antonelli, G. (2018). *Underwater Robots*. Springer. https://doi.org/10.1007/978-3-319-72124-8

Bahr, A., Leonard, J. J., & Fallon, M. (2009). Cooperative Localization for Autonomous Underwater Vehicles. *The International Journal of Robotics Research*, *28*, 714–728.

Balasuriya, A., & Ura, T. (2002). Autonomous Target Tracking by Underwater Robots Based on Vision. In *Proceedings of 1998 International Symposium on Underwater Technology* (vol. *29*, pp. 1351–1367). IEEE. https://doi.org/10.1109/ut.1998.670089

Carder, K., Costello, D., Warrior, H., Langebrake, L., Hou, W., Patten, J., & Kaltenbacher, E. (2001, October). Ocean-science Mission Needs: Real-time AUV Data for Command, Control, and Model Inputs. *IEEE Journal of Oceanic Engineering, 26*, 742–751. https://doi.org/10.1109/48.972116

Cecchi, D., Caiti, A., Fioravanti, S., Baralli, F., & Bovio, E. (2005). Target Detection Using Multiple Autonomous Underwater Vehicles. In *Proceedings IARP International Workshop on Underwater Robotics* (pp. 161–168). Genova.

Chaudhary, M., Goyal, N., Benslimane, A., Awasthi, L., Alwadain, A., & Singh, A. (2023). Underwater Wireless Sensor Networks: Enabling Technologies for Node Deployment and Data Collection Challenges. *IEEE Internet Things Journal, 10*, 3500–3524.

Costanzi, R., Manzari, V., Teracciano, D., Tesei, A., & Fenucci, D. (2020, July). Interoperability Among Unmanned Maritime Vehicles: Review and First In-field Experimentation. *Frontiers of Robotics and AI* (7), 1–15.

Defense Science Board. (2015). *Next-Generation Unmanned Undersea Systems.* Technical report, Defense Science Board, U.S. Department of Defense.

Finn, A., & Scheding, S. (Eds.). (2010). *Developments and Challenges for Autonomous Unmanned Vehicles: A Compendium.* Springer-Verlag Berlin Heidelberg. https://doi.org/10.1007/978-3-642-10704-7

Ghabcheloo, R., Pascoal, A., Silvestre, C., & Kaminer, I. (2009). Coordinated Path Following Control of Multiple Vehicles Subject to Bidirectional Communication Constraints. In *Lecture Notes in Control and Information Science* (pp. 93–111). Seoul: Springer-Verlag. https://doi.org/10.1007/11505532_6

Griffiths, G. (Ed.). (2003). *Technology and Applications of Autonomous Underwater Vehicles.* Taylor & Francis.

Gupta, S., & Singh, N. (2017). Underwater Wireless Sensor Networks: A Review of Routing Protocols, Taxonomy, and Future Directions. *The Journal of Supercomputing, 80,* 5163–5196. https://doi.org/10.1007/s11227-023-05646-w

Hasan, H. (2017). *Underwater Robotics: Science, Design and Fabrication.* Malaysia: UTM Press: Malaysia.

Heshmat, M., Saoud, L. S., Abujabal, M., Sultan, A., Elmezain, M., Seneviratne, L., & Hussain, I. (2025). Underwater SLAM Meets Deep Learning: Challenges, Multi-Sensor Integration, and Future Directions. *Sensors, 25*(11). https://doi.org/10.3390/s25113258

Hwangbo, S.-H., Jeon, J.-H., & Park, S.-J. (2013). Wireless Underwater Monitoring Systems Based on Energy Harvestings. *Sensors & Transducers, 18*, 113–119. https://www.researchgate.net/publication/290078574_Wireless_Underwater_Monitoring_Systems_Based_on_Energy_Harvestings

Hyland, B. (2016, October). Subsea Internet of Things to Transform Amount of Information Available. *Hydro International, 20.* www.hydro-international.com

Inzartsev, A., & Pavi, A. (2009, January 1). AUV Application for Inspection of Underwater Communications. In A. V. Inzartsev (Ed.), *Underwater Vehicles* (pp. 371–398). InTech. https://doi.org/10.5772/6704

Khan, N., Kalair, A., Abas, N., & Haider, A. (2017, May). Review of Ocean Tidal, Wave and Thermal Energy Technologies. *Renewable and Sustainable Energy Reviews, 72,* 590–604.

Kim, H., Lee, Y., & Myung, H. (2020). Factor Graph Optimization for Underwater Navigation with Acoustic Signals. *IEEE Access, 8*, 77720–77731.

Leonard, J., & Bahr, A. (2016). Autonomous Underwater Vehicle Navigation. In *Springer Handbook of Ocean Engineering* (pp. 341–358). Springer.

Liu, J., Yu, F., He, B., & Soares, C. G. (2024). A Review of Underwater Docking and Charging Technology for Autonomous Vehicles. *Ocean Engineering*, *297*(1), 117–154.

Mohsan, S. A., Khan, M. A., Mazinani, A., Alsharif, M. H., & Cho, H.-S. (2022). Enabling Underwater Wireless Power Transfer towards Sixth Generation (6G) Wireless Networks: Opportunities, Recent Advances, and Technical Challenges. *Journal of Marine Science and Engineering, 10*(9), 1–36.

Nichols, R. K., Mumm, H. C., Ryan, J., Lonstein, W., Carter, C., Hood, J. P., . . . Shay, J. (2020). *Unmanned Vehicle Systems & Operations on Air, Sea, Land.* New Prairie Press. https://doi.org/10.4148/2334-0402.1030

Nkenyereye, L., Nkenyereye, L., & Ndibanje, B. (2024). Internet of Underwater Things: A Survey on Simulation Tools and 5G-Based Underwater Networks. *Electronics (Special Issue: Artificial Intelligence Empowered Internet of Things), 13*(3), 1–41.

PIB. (2025, November 14). *DRDO Develops New Generation Man-portable Autonomous Underwater Vehicles for Mine Countermeasure Missions.* Press Information Bureau. https://www.pib.gov.in/PressReleasePage. aspx?PRID=2190096

Qiu, T., Zhao, Z., Zhang, T., Chen, C., & Chen, C. (2020). Underwater Internet of Things in Smart Ocean: System Architecture and Open Issues. *IEEE Transactions on Industrial Informatics, 16*, 4297–4307.

Randeni, S., Schneider, T., Bhatt, E. C., Víquez, O. A., & Schmidt, H. (2022, November). A High-Resolution AUV Navigation Framework with Integrated Communication and Tracking for Under-ice Deployments. *Journal of Field Robotics, 40*(2), 346–367. https://doi.org/10.1002/rob.22133

Rice, J. A., & Green, D. (2008). Underwater Acoustic Communications and Networks for the US Navy's Seaweb Program. *Sensor Technologies and Applications SENSORCOMM '08.* San Diego. https://doi.org/10.1109/SENSORCOMM.2008.137

Roberts, G. N., & Sutton, R. (Eds.). (2013). *Further Advances in Unmanned Marine Vehicles.* The Institution of Engineering and Technology.

Stommel, H. (1989, April). The Slocum Mission. *Oceanography.*

Teixeira, F. C., & Pascoal, A. M. (2007). Geophysical Navigation of Autonomous Underwater Vehicles. *IFAC Proceedings, 40*(17), 117–122. https://doi.org/10.3182/20070919-3-HR-3904.00022

Wadoo, S., & Kachroo, P. (2011). *Autonomous Underwater Vehicles: Modeling, Control, Design, and Simulation.* CRC Press Taylor & Francis Group.

Wang, D., Zhang, J., Cui, S., Bie, Z., Chen, F., & Zhu, C. (2024). The State-of-the-arts of Underwater Wireless Power Transfer: A Comprehensive Review and New Perspectives. *Renewable and Sustainable Energy Reviews, 189*(A).

Yang An, G. W. (2025, November). A Study on the Performance of a Novel Underwater Current Energy Harvester in Environments with Obstacles. *Energy, 336*, 1–30.

Yuan, X., Martínez, J.-F., Pons, J. S., & Eckert, M. (2017, May). AEKF-SLAM: A New Algorithm for Robotic Underwater Navigation. *Sensors, 17*(5), 1174. https://doi.org/10.3390/s17051174

Zeng, F., & Wang, T. (2023, October). In-situ Wave Energy Harvesting for Unmanned Marine Devices: A Review. *Ocean Engineering, 285,* 1–41.

Zhang, S., Zhao, S., An, D., Liu, J., Wang, H., Feng, Y., . . . Zhao, R. (2022, November). Visual SLAM for Underwater Vehicles: A Survey. *Computer Science Review, 46.* https://doi.org/10.1016/j.cosrev.2022.100510

Zuluaga, C., Aristizábal, L., Rúa, S., Franco, D., Osorio, D., & Vásquez, R. (2022). Development of a Modular Software Architecture for Underwater Vehicles Using Systems Engineering. *Journal of Marine Science and Engineering, 10*, 464.

Index

For Product Safety Concerns and Information please contact our EU representative GPSR@taylorandfrancis.com Taylor & Francis Verlag GmbH, Kaufingerstraße 24, 80331 München, Germany

Batch number: 10725802

Printed by Printforce, the Netherlands